**LHS★**

**Great Explorations in Math & Science (GEMS)**

**Lawrence Hall of Science**
University of California
Berkeley, California 94720-5200
510 642-7771 Fax: 510 643-0309

 W9-BDL-226

May 7, 2001

Dear Trial Test Teacher,

At long last, the final version of *Environmental Detectives* has been published!

On behalf of the Lawrence Hall of Science and the Great Explorations in Math and Science program, we thank you again for taking time to field test our teacher's guides and for providing us with your feedback.  Your feedback serves to greatly improve the overall usefulness of the guides, and is an invaluable contribution to their educational value.

Enclosed is a complimentary copy of *Environmental Detectives*. Please let us know what you think!  Don't forget to check the front section of the guide to confirm that your name and your school's name were properly credited.

Best regards,

Cheryl Webb
Trial Test Coordinator

cwebb@uclink4.berkeley.edu
(510) 643-1129

# Environmental Detectives

## Teacher's Guide

### Grades 5–8

### Skills
Observing, Comparing, Inferring, Researching, Visualizing, Explaining, Analyzing and Evaluating Evidence, Making Inferences, Distinguishing Evidence from Inference, Recording, Organizing Data, Working Cooperatively, Communicating, Logical Thinking, Drawing Conclusions, Problem Solving

### Concepts
Aquatic Biology, Fish Habitats, Pollution, Point and Non-Point Polluters, Acid Rain, Water Contamination, Erosion, Sediments, Chlorine Tests, Acid/Base Chemistry, Predator/Prey, Animal Population Dynamics, Human Impact on the Environment

### Themes
Models and Simulations, Systems and Interactions, Patterns of Change

### Mathematics Strands
Number, Measurement, Pattern, Statistics, Logic and Language

### Nature of Science and Mathematics
Scientific Community, Cooperative Efforts, Creativity and Constraints, Interdisciplinary, Real-Life Applications

*by*
**Kevin Beals** *with* **Carolyn Willard**

**LHS GEMS**

**Great Explorations in Math and Science**
**Lawrence Hall of Science**
**University of California at Berkeley**

**Cover Design**
Lisa Klofkorn
**Design and Illustrations**
Lisa Klofkorn

**Photographs**
Richard Hoyt
Laurence Bradley

Lawrence Hall of Science, University of California,
Berkeley, CA 94720-5200

**Director:** Ian Carmichael

Initial support for the origination and publication of the GEMS series was provided by the A.W. Mellon Foundation and the Carnegie Corporation of New York. Under a grant from the National Science Foundation, GEMS Leader's Workshops have been held across the country. GEMS has also received support from: the McDonnell-Douglas Foundation and the McDonnell-Douglas Employee's Community Fund; Employees Community Fund of Boeing California and the Boeing Corporation; the Hewlett Packard Company; the people at Chevron USA; the William K. Holt Foundation; Join Hands, the Health and Safety Educational Alliance; the Microscopy Society of America (MSA); the Shell Oil Company Foundation; and the Crail-Johnson Foundation. GEMS also gratefully acknowledges the contribution of word processing equipment from Apple Computer, Inc. This support does not imply responsibility for statements or views expressed in publications of the GEMS program. For further information on GEMS leadership opportunities, or to receive a catalog and the *GEMS Network News*, please contact GEMS at the address and phone number below. We also welcome letters to the *GEMS Network News*.

Printed on recycled paper with soy-based inks.

International Standard Book Number: 0-924886-23-4

## COMMENTS WELCOME !

Great Explorations in Math and Science (GEMS) is an ongoing curriculum development project. GEMS guides are revised periodically, to incorporate teacher comments and new approaches. We welcome your criticisms, suggestions, helpful hints, and any anecdotes about your experience presenting GEMS activities. Your suggestions will be reviewed each time a GEMS guide is revised. Please send your comments to: GEMS Revisions, c/o Lawrence Hall of Science, University of California, Berkeley, CA 94720-5200. The phone number is (510) 642-7771 and the fax number is (510) 643-0309. You can also reach us by e-mail at gems@uclink4.berkeley.edu or visit our web site at www.lhsgems.org.

# Great Explorations in Math and Science (GEMS) Program

The Lawrence Hall of Science (LHS) is a public science center on the University of California at Berkeley campus. LHS offers a full program of activities for the public, including workshops and classes, exhibits, films, lectures, and special events. LHS is also a center for teacher education and curriculum research and development.

Over the years, LHS staff have developed a multitude of activities, assembly programs, classes, and interactive exhibits. These programs have proven to be successful at the Hall and should be useful to schools, other science centers, museums, and community groups. A number of these guided-discovery activities have been published under the Great Explorations in Math and Science (GEMS) title, after an extensive refinement and adaptation process that includes classroom testing of trial versions, modifications to ensure the use of easy-to-obtain materials, with carefully written and edited step-by-step instructions and background information to allow presentation by teachers without special background in mathematics or science.

## Staff

**Director:** Jacqueline Barber
**Associate Director:** Kimi Hosoume
**Associate Director/Principal Editor:** Lincoln Bergman
**Mathematics Curriculum Specialist:** Jaine Kopp
**GEMS Network Director:** Carolyn Willard
**GEMS Workshop Coordinator:** Laura Tucker
**Staff Development Specialists:** Lynn Barakos, Katharine Barrett, Kevin Beals, Ellen Blinderman, Gigi Dornfest, John Erickson, Stan Fukunaga, Philip Gonsalves, Debra Harper, Linda Lipner, Karen Ostlund
**Distribution Coordinator:** Karen Milligan
**Workshop Administrator:** Terry Cort
**Financial Assistant:** Vivian Tong

**Distribution Representative:** Felicia Roston
**Shipping Assistant:** Maureen Johnson
**Director of Marketing and Promotion:** Matthew Osborn
**Trial Test and Materials Manager:** Cheryl Webb
**Senior Editor:** Carl Babcock
**Editor:** Florence Stone
**Principal Publications Coordinator:** Kay Fairwell
**Art Director:** Lisa Haderlie Baker
**Senior Artists:** Carol Bevilacqua, Rose Craig, Lisa Klofkorn
**Staff Assistants:** Trina Huynh, Thania Sanchez, Dareyn Stilwell, Stacey Touson, Jennifer Yee

## Contributing Authors

Jacqueline Barber
Katharine Barrett
Kevin Beals
Lincoln Bergman
Susan Brady
Beverly Braxton
Kevin Cuff

Linda De Lucchi
Gigi Dornfest
Jean Echols
John Erickson
Philip Gonsalves
Jan M. Goodman
Alan Gould

Catherine Halversen
Debra Harper
Kimi Hosoume
Susan Jagoda
Jaine Kopp
Linda Lipner
Larry Malone

Cary I. Sneider
Craig Strang
Herbert Thier
Jennifer Meux White
Carolyn Willard

# *Reviewers*

We would like to thank the following educators who reviewed, tested, or coordinated the reviewing of this group of GEMS guides *(Ocean Currents, Only One Ocean,* and *Environmental Detectives.)* Their critical comments and recommendations, based on classroom presentation of these activities nationwide, contributed significantly to this publication. Their participation in this review process does not necessarily imply endorsement of the GEMS program or responsibility for statements or views expressed. Their role is an invaluable one; feedback is carefully recorded and integrated as appropriate into the publications. **THANK YOU!**

## ARIZONA

**Sonoran Sky Elementary School, Scottsdale**
Brent Engilman
Marge Maceno *
Amy Smith
Kathy Wieeke
Tammy Wopnford

## ARKANSAS

**Bob Courtway Middle School, Conway**
Robin Cole
Rick Hawkins
Charlcie Strange *
Paula Wilson

**Birch Kirksey Middle School, Rogers**
Beth Ann Carnes
Jenny Jones *
Curtis S. Smith
Sharron Wolf

## CALIFORNIA

**Albany Middle School, Albany**
Karen Adams
Jenny Anderson
Cyndy Plambeck
Kay Sorg

**Martin Luther King Middle School, Berkeley**
Indigo Babtiste
Akemi Hamai
Yvette McCullough
Beth Sonnenberg

**Harding School, El Cerrito**
Renie Gannett
Carol Leitch
Jim Wright

**Portola Middle School, El Cerrito**
Debbie Marasaki
Carol Mitchell
Susan Peterson
Mike Wilson

**Warwick Elementary School, Fremont**
Dale Harden
Katy Johnson
Richard Nancee
Robert Nishiyam
Bonnie Quigley
Ann Trammel

**Ohlone Elementary School, Hercules**
Stacey Cragholm
Gloria Crim
Jay Glesener
Sandra Simmons

**Hall Middle School, Larkspur**
Trish Mihalek
Art Nelson
Ted Stoeckley
Barry Sullivan

**McAuliffe Middle School, Los Alamitos**
Michelle Armstrong
Kathy Burtner

**Oak Middle School, Los Alamitos**
Joyce Buehler
Rob Main *
Kendall Vaught

**Hidden Valley School, Martinez**
Diane Coventry
Nigel Dabby
Jennifer Sullivan

**Calvin Simmons Jr. High School, Oakland**
Stan Lake
Wendy Lewis
Fernando Mendez
Thelma Rodriguez

**Vintage Parkway School, Oakley**
Jennifer Asmussen
Alisa Haley
Casey Maupin
Lian McCain
Steve Williams

**Collins Elementary School, Pinole**
Ralph Baum
Craig Payne
Anne Taylor
Genevieve Webb

**Adams Middle School, Richmond**
Richard Avalos
Susan Berry
John Eby
John Iwawaki
Steve Stewart

**Bell Junior High School, San Diego**
Nick Kardouche
Elouise King *
Denise Vizcarra
Mala Wingerd

**Dingeman Elementary, San Diego**
Monka Ely *
Godwin Higa
Kim Holzman
Linda Koravos

**Cook Middle School, Santa Rosa**
Steve Williams

**Rincon Valley Middle School, Santa Rosa**
Sue Lunsford
Penny Sirota *
Laurel VarnBuhler

## FLORIDA

**Howard Middle School, Orlando**
Elizabeth Black
Carletta Davis
Susan Leeds *
Jennifer Miller

## MISSOURI

**St. Bernadette School, Kansas City**
Brett Coffman
Dorothy McClung
Aggie Rieger
Margie St. Germain

**Poplar Bluff 5th/6th Grade Center, Poplar Bluff**
Cindy Gaebler
Leslie Kidwell
Barbara King *
Melodie Summers

## NEVADA

**Churchill County Jr. High, Fallon**
Kerri Angel
Deana Madrasco
Amy Piazzola
Sue Smith-Ansotegui *

## NEW HAMPSHIRE

**Crescent Lake Elementary, Wolfeboro**
Kate Borelli
Amy Kathan
Elaine M. Meyers *
Patti Morissey

## NEW JERSEY

**Orchard Hill School, Skillman**
Jay Glassman *
Al Hadinger
Georgiana Kichura
Tony Tedesco

## NEW YORK

**Maple Hill Middle School, Castleton**
Beth Chittendo
Jeanne Monteau *

**Lewisboro Elementary School, South Salem**
Debra Jeffers

**St Brigid's Regional Catholic School, Wateruliet**
Patricia Moyles

## OHIO

**Baker Middle School, Marion**
Dave Dotson
Denise Z. Iams *
Betty Oyster
Carol White

## OREGON

**Sitton School, Portland**
David Lifton
Deborah Nass

## TEXAS

**Colleyville ISD, Grapevine**
Kathy Keeney

**Grapevine ISD / Admin. Bldg., Grapevine**
Shelly Castleberry
Terry Dixon
Malanie Gable*
Randy Stuempfig

**Ector County ISD, Odessa**
Becky Stanford

**Ireland Elementary, Odessa**
Susan Hardy

**Miliam Elementary, Odessa**
Eli Tavarez

**Travis Elementary, Odessa**
Patty Calk
Stacey Hawkins

## WASHINGTON

**The Gardner School, Vancouver**
Matt Karlsen
Tom Schlotfeldt
Rob VanNood

* Trial test coordinators

# Acknowledgments

The mystery framework for this guide and many of the aspects of the environmental situation it depicts are derived from successful Lawrence Hall of Science outreach programs. In this GEMS guide, rather than making use of the learning station approach (also used in the GEMS *Mystery Festival* guide) we have chosen instead to present major topics in more depth while retaining the compelling nature of the mystery framework.

The original idea for an environmental festival came from former LHS staff member (and current *Grossology* author) Sylvia Branzei. Other LHS staff who worked on the original festival include Lynn Barakos, Jacquey Barber, Kevin Beals, Anne Brocchini, Jennifer Claesgens, Laura Lowell, Mayumi Shinohara, Rebecca Tilley, and Lisa Walenceus. Elements of the festival were also inspired by the book *Who Really Killed Cock Robin?* by Jean Craighead George.

The "Deer Lion" dice game in Activity 5 and the more active alternate version were originally inspired by the "Oh Deer" game in the educational materials of Project WILD. Modifications and adaptations were made by the authors and the game was further modified during the GEMS trial test process.

The teachers who took part in testing this unit are listed by region and school in the front of this guide. We much appreciate their efforts! Tom LaHue of Aptos, California, and several teachers from the GEMS Network Site in Merrillville, Indiana, were eager to try out these activities on their own, creating their own kits to do so. Their feedback also provided valuable insight as we revised the activities for publication. Very special thanks to Marcelino Echeverria for portraying Juan Tunó. In addition to co-author Carolyn Willard, other GEMS and LHS staff who provided valuable input include Jacqueline Barber, Kimi Hosoume, Lincoln Bergman, and Florence Stone.

# Contents

"All things are connected…"

—Chief Sealth (also known as Chief Seattle)

"When we try to pick out anything by itself, we find that it is bound fast by a thousand invisible cords that cannot be broken to everything in the universe."

—John Muir

# Introduction

Who doesn't enjoy the fun and challenge of trying to solve a mystery? Scientific investigation is about gathering evidence toward the solution of the mysteries of the natural world. This guide takes the mystery "hook," successfully featured in such GEMS guides as *Mystery Festival, Crime Lab Chemistry, Fingerprinting,* and one portion of *Learning About Learning*, and applies it to the environment.

In this case, the "scene of the crime" is the "Gray Area," a watershed that includes forests, a city and town, a coast, three rivers, a lake, and a pond. The "crime" is a fish die-off that began five years ago. As students learn more about the possible causes of the fish dying, the complexity of the mystery mounts. They learn how much "gray area" there is in science—and how difficult it can be to pinpoint the exact cause of an environmental problem. They become aware of the interconnectedness of the natural world and of environmental problems, and see how one small change can trigger a whole chain of events. Students learn how science and society are inextricably linked, and that most solutions require compromise. They discover that environmental problems are not only caused by "big bad companies," but also by the general public, including themselves.

The unit is not a series of laboratory tests. Although students *do* conduct some simulated environmental tests, the real value of *Environmental Detectives* is that students grapple with a complex, interdisciplinary scientific investigation. Students hear statements of various "suspects" in the "crime." They study and discuss reference materials, including records, newspaper articles, charts, graphs, and even "secret documents," and integrate all of this information with their own test results. Frequent discussions provide opportunities to process information and apply concepts. Following each small group discussion is an all class teacher-led discussion, in which the students put together all the information gained that day. Each activity concludes with an opportunity for students to reflect, to write in their Environmental Detective Notebooks, and to make predictions or adjust previous predictions on the class Suspect Chart. The notebooks can also be used for assessment purposes. Optional homework question sheets are included, to encourage students to process information on their own. These sheets can also help you find out what is going on in each team's discussions and can be used for assessment purposes.

There are many environmental problems worthy of study, but we've chosen to focus on problems students can relate to, as well as those they can impact through the choices they make in their own lives. These problems include: chlorine pollution, acid rain, erosion/sediment pollution, predator-prey relationships, phosphate pollution/algal blooms, and oil pollution. Of course, each of these would take a long time to study thoroughly. In this unit, students learn enough about a variety of these problems to understand how interconnected they may be in a real-life situation.

Students are highly motivated to solve the mystery, and the unit is designed so, as much as possible, information and ideas come from the students themselves, rather than from the teacher. **It's important to be patient and let students discover information for themselves.** Specific portions or bits of information in the mystery may elude certain students or classes, but with so much content involved, the big ideas are most important. It is not crucial that students fully understand every aspect of each topic.

A brief reassuring word about preparation and presentation time is warranted (also see "Time Frame" on page 7). This is a substantial unit, providing in-depth instruction in a number of major content areas. This unit can be aligned with national standards, places science in a relevant real-world context, and is highly motivating. Preparation tasks and duplication of student sheets are also substantial. As with several other similar GEMS guides, parents and other assistants can be very helpful. Many of the duplicated sheets can be used again and again as the activities are presented to other classes. We have offered suggestions for streamlining preparation whenever possible, and are sure you will come up with more! Let us hear from you. The message we heard from the teachers who tested this guide in their classrooms nationwide came through loud and clear—it is well worth the effort!

*One trial test teacher wrote: "This unit is an outstanding science investigation for middle school. It gives students a purpose for learning certain science concepts and skills. It allows them to be active, reflective learners who must learn certain information to solve a problem."*

## National Content Standards and Guided Inquiry

As the first page of this guide summarizes, these activities bring forth a great deal of scientific—and mathematical—learning. The major content areas students investigate can be closely correlated to the *National Science Education Standards,* as well as to state science frameworks and district guidelines. The unit interweaves chemistry, the life sciences, and earth science with environmental science. It connects strongly to what the *National Standards* call "science and technology" and "science in personal and social perspectives," which include relevant sub-categories, such as "understandings about science and technology," "populations, resources, and environments," "natural hazards," "risks and benefits," and "science and technology in society."

The *National Standards* also place strong emphasis on the "science and inquiry" standard, which relates to developing both student inquiry abilities and student understanding of scientific inquiry/investigation/methods. In its overall framework—the investigation and solution of an environmental problem—this unit definitely models a many-sided scientific investigation. *Environmental Detectives* is an especially excellent way to provide students with practice in a key and invaluable aspect of scientific inquiry—doing research. Most students need encouragement, practice, and guidance to develop and refine their inquiry abilities in general, and their research skills in particular. This unit will help them learn to appreciate the power of good research.

The research students conduct in the *Environmental Detectives* "files" provides many opportunities to improve their skills and deepen their understanding. They get practice combing through multiple documents, deciding which are pertinent, and then gleaning the information they need. Students, like other researchers, may even find they can't resist reading interesting documents, even if they seem not to be directly useful. You never know when that odd tidbit of knowledge will come in handy!

Your students are likely to find that the research comes more easily with each successive activity in the unit. Students also learn that teamwork can make research easier, and that group discussion can lead to greater insights. With its emphasis on research (and its overall structure for a scientific investigation) this GEMS unit makes a contribution to the development of student inquiry abilities.

In testimony to a Congressional education committee in late 1999, Dr. Herbert Thier of Lawrence Hall of Science, Director of the Science Education for Public Understanding Program (SEPUP) discussed "guided inquiry" as part of a new, context-centered view of science education reflected in the *National Science Education Standards*. In this excerpt, he could have been talking about this unit: "In guided inquiry, processes and information are integrated and sequenced to guide students as they confront scientific concepts and principles in the context of real-world problems. Students are expected to use data and evidence to reason their way through a particular problem or issue; and reach independent conclusions or decisions justified by the data and evidence. Placing scientific ideas and processes in the context of actual issues—balancing the risk and benefits of genetically engineered food, for example, or of industrial production—can suddenly give formerly abstract concepts meaning within students' own lives, a key element in helping them master knowledge."

## Activity-by-Activity Overview

### Setting the Scene

Some wall or chalkboard space is necessary to set up "Headquarters." This area consists of a large map of the Gray Area, and a chart of "suspect" pictures and statements, on which students place post-its to make predictions of who is "guilty." Also stored near Headquarters are the files that students will use to do their research throughout the unit.

Most of the eight main activities have a similar overall pattern—introduce the topic, perform tests or learn of data gathered by others, do research on files in groups of four, discuss in groups of four, discuss with the whole class, and finally, reflect, change predictions, and record in notebooks. This pattern repeats somewhat so students (and teachers) are comfortable with the procedure and know what to expect from day to day. However, to prevent the process from becoming too tedious, the formats of the individual activities do vary and surprises are scattered here and there.

### Main Activities

In Activity 1, the students are introduced to the Gray Area and its problem, then begin brainstorming information about the needs of fish and ideas as to what might be causing the fish kill. They then use Timeline Clue Cards to learn about the history of the area, and start their Environmental Detective Notebooks. The timeline will be useful later in the unit as they piece together clues about alleged polluters. They meet three characters from the Gray Area, and hear their statements.

In Activity 2, students learn about possible chlorine pollution in the area, related to the local water slide. Through a role play of a meeting of the Gray Area Board of Supervisors, they receive the results of chemical and biological tests for chlorine. Teams of four do research in the Chlorine Files. During small and large group discussions, they attempt to resolve the seemingly contradictory results from the two tests. The tests in this activity are designed to be paper and pencil simulations, but if possible, **we strongly recommend that you use real daphnia instead.** Information on how to obtain and set up daphnia samples can be found on page 222.

In Activity 3, students take on the problem of acid rain. At hands-on learning stations around the room, they perform pH tests on local waterways. Once they've determined which waterways have a problem, they attempt to figure out the source of the problem. Because of the rich activity, research in files and discussion, the **Acid Rain activity usually takes two class sessions.**

In Activity 4, erosion and sedimentation are introduced and modeled through a teacher demonstration on the overhead projector. Pairs of students then do hands-on tests at eight learning stations to determine the amount of sediment in different water sites from the Gray Area. Again, students study and discuss their results, plus the contents of the Sediment Files, hold a class discussion, and record the latest information in their notebooks.

In Activity 5, students are introduced to the idea that changes in the populations of deer may be causing erosion and increased sedimentation in a local river. They play a game simulating deer population dynamics in which they graph the populations of deer and mountain lions over a series of "years." After the simulation they study the graph to notice trends, and determine the effect the mountain lions had upon the deer population. **Some teachers prefer to do this activity in two class sessions.**

In Activity 6, students are introduced to the phosphate and algal bloom problem at James Pond. They perform two simulated field tests (birdwatching and water life), and a hands-on test for phosphates in the James Pond area. Including the tests themselves, and research, discussion, reflection and recording, **the James Pond activity usually takes two class sessions.**

In Activity 7, rather than conducting their own tests and research, students find out about oil pollution through testimony at a board meeting. The students examine data to identify the type of oil found in Gray Bay. After a class discussion of all the suspected problems introduced in the entire unit, they vote, individually and as a group of four, to decide which factors they think are most responsible for the fish kill.

In Activity 8, students change from the role of environmental detectives into that of a local decision-making board. They brainstorm solutions to the problems they concluded were most likely to be killing the fish. They then debate the pros and cons of each solution, and vote on whether or not to pass it. The collection of passed solutions then becomes their environmental policy.

## Connecting to Real Life

As implied by all of the above, *Environmental Detectives* places science learning in a real-world context—with issues like those we hear about every day. The "mystery" scenario adds a further level of interest and student motivation, and students are empowered to seek their best solutions to complex environmental problems. Along the way, they find out about real tests environmental scientists conduct. They work with and analyze statistical and other data, just as scientists (and many other professions) must do.

It's interesting that the way the full-scale investigation begins is triggered by a student, Juan Tunó, taking scientific matters into his own hands because he does not think the water slide—a popular recreation for young people—is responsible for the fish dying. Of course your students will relate to this youthful impetus—it is also well worth emphasizing that many young students all over the world—girls and boys— have played an important role in calling attention to and proposing solutions for environmental issues. A number of related resources will be listed in the guide, such as *Kid Heroes of the Environment*, that your students could read to find out more about these real-life inquirers and problem solvers!

As with all GEMS teacher's guides, we include "Assessment Suggestions" to highlight those activities and assignments that particularly lend themselves to finding out how well students have learned central concepts. "Literature Connections" focus on age-appropriate books that connect to or extend big ideas in the guide. "Summary Outlines" (including preparation steps) are included to assist you in classroom presentation.

That said, it's time for you to tell your students to get ready, put on their thinking caps, and travel to the land of the Missterssippi River, where they will get to be "environmental detectives!"

# Time Frame

Depending on the age and experience of your students, the length of your class periods, and your teaching style, the time needed for the *Environmental Detectives* unit may vary. Try to build flexibility into your schedule so that you can extend the number of class sessions if necessary.

Activities 3, 5, and 6 will definitely take two sessions each to complete, and you may decide to allow even more time for additional student discussion and reflection. In other activities, we have indicated good places to break if two sessions are needed.

Activity 1: Introducing the Mystery ........... one or two 45–60 minute sessions

Activity 2: Chlorine Tests ............................ one or two 45–60 minute sessions

Activity 3: Acid Rain .................................... two 45–60 minute sessions

Activity 4: Sedimental Journey ................................. one 45–60 minute session

Activity 5: Deer Lion ................................. two or three 45–60 minute sessions

Activity 6: James Pond Tests ................................. two 45–60 minute sessions

Activity 7: Oil and "Who Done It?" ......................... one 45–60 minute session

Activity 8: Solving the Problems .............................. one 45–60 minute session

**Special Note:**
Although the number of class sessions it takes to complete this unit may seem like a lot, in some ways it is quite an efficient use of time. Please see the main introduction for a summary of how this unit meets learning objectives in many disciplines of science, while promoting inquiry/research skills and logical thinking abilities. Also consider the interdisciplinary nature of the unit, and its strong cross-curricular emphases on reading, language arts, the environment, and social studies! And yes, there is a great deal of paper to duplicate, but most of these sheets are "non-consumable" and can be used again by multiple classes. This and other preparation, while substantial, becomes well worth it as your students start to grapple with the rich learning "environment" of this unit!

# What You Need for the Whole Unit

The quantities below are based on a class size of 32 students. Depending on the number of students in your class, you may, of course, need different amounts of materials.

This list gives you a concise "shopping list" for the entire unit. Please refer to the "Getting Ready" sections for each activity. They contain more specific information about the materials needed for the class and for each team of students.

Many photocopies are necessary for this unit. You may want to enlist a volunteer to do the copying. Also, the copying could be done as needed for each activity, or all the necessary pages could be copied at one time. Further, the copied pages could be organized by using color paper. For example, the clue cards, discussion cards, and files for an activity could all be copied on one color paper. This would make it easier to return any stray papers to their proper location.

## Non-Consumables

- ❐ 4 copies of the record of the Meeting of the Gray Area Board of Supervisors (master on pages 43–44)
- ❐ 1 copy of the Water Flea Diagram (master on page 42)
- ❐ 9 plastic cups (8–10 oz. size)
- ❐ a cottage cheese type or other container (about 2 cup size)
- ❐ 3 copies of the pH Scale (master on page 77)
- ❐ 2 two-liter clear plastic soda bottles
- ❐ a chalkboard, butcher paper, or overhead transparency copy of the Acid Rain Questions (master on page 80)
- ❐ several small paper bags or garbage cans
- ❐ 1 or 2 sponges for cleaning up spills
- ❐ 1 *transparent* glass or plastic tray or pan at least 9" x 12" and at least 2" deep. (A sturdy container is preferable. If you use a lightweight container, such as a flat deli-style salad container, be sure to test it before class to make certain it doesn't leak or tip over easily.)
- ❐ 2 plastic flat-bottomed cups (about 8 or 9 oz. size)

- ❏ a small bowl or cup for extra water
- ❏ 1 tablespoon measuring spoon or a graduated cylinder (15 ml or more)
- ❏ 8 large washers
- ❏ 8 one-foot pieces of white string
- ❏ 8 milk cartons (half-gallon or quart size)
- ❏ spoon or ruler
- ❏ 3 reaction trays with at least 3 compartments OR 3 sets of 3 plastic cups
- ❏ 3 medicine droppers OR plastic straws cut in half to use as droppers (see "Getting Ready," page 145)
- ❏ access to a sink, or a dishtub with a squirt bottle of water and a sponge
- ❏ 1 copy of the What Happens in an Algal Bloom? pages (masters on pages 161–164)
- ❏ 4 copies of James Pond (master on page 167)
- ❏ 4 copies of the Water Life Identification Key (master on page 168)
- ❏ 8 copies of the Bird Identification Card (master on page 170)
- ❏ 16 copies of the Birdwatching at James Pond sheets (masters on pages 171–172)
- ❏ 6 copies of the record of the Emergency Meeting of the Gray Area Board of Supervisors (master on pages 197–198)

An overhead transparency of each of the following:
- ❏ Map of the Gray Area (master on page 25)
- ❏ Gray Area History Timeline (master on page 26)
- ❏ Deer and Mountain Lion Populations in Parallel Park (master on page 134)
- ❏ Oil Chromatography Test Data (master on page 195)
- ❏ Fish Autopsy Results (master on page 196)

One copy of each of the following pictures/statements:
- ❏ Juan Tunó's picture and Introductory statement (master on page 27)
- ❏ Don Juan Tunó's picture and Introductory statement (master on page 28)
- ❏ Avery Wun's picture and Introductory statement (master on page 29)
- ❏ Juan Tunó's Chlorine statement (master on page 40)
- ❏ Ken Unballe's picture and Chlorine statement (master on page 41)
- ❏ LaToya Faktorie's picture and Acid Rain statement (master on page 72)
- ❏ Juan Tunó's Acid Rain statement (master on page 71)

❑ Don Juan Tunó's Acid Rain statement (master on page 73)

❑ Anton Alogue's picture and Sediment statement (master on page 104)

❑ Elmo Skeeto's picture and Sediment statement (master on page 105)

❑ Juan Tunó's Sediment statement (master on page 103)

❑ Juan Tunó's Deer Lion statement (master on page 131)

❑ Bo Vyne's picture and James Pond statement (master on page 159)

❑ Sandy Trapp's picture and James Pond statement (master on page 160)

❑ Juan Tunó's James Pond statement (master on page 157)

❑ Don Juan Tunó's James Pond statement (master on page 158)

❑ Final Suspect Statements (masters on pages 199–201)

❑ Mandy Lyfbotes' picture and Oil statement (master on page 194)

Eight sets of the following clue cards:

❑ Timeline Clue Cards (masters on pages 30–31)

❑ Acid Rain Clue Cards (masters on pages 82–85)

Eight copies of the following files:

❑ Chlorine Files (masters on pages 45–49)

❑ Acid Rain Files (masters on pages 86–88)

❑ Sediment Files (masters on pages 107–110)

❑ Deer Lion Files (masters on pages 134–138)

❑ James Pond Files (masters on pages 177–181)

Eight each of the following discussion cards:

❑ Chlorine Discussion Card (master on page 50)

❑ Deer Lion Discussion Card (master on page 139)

❑ James Pond Discussion Card (master on page 182)

The following procedure sheets:

❑ 8 copies of the pH Procedure Sheet (master on page 74)

❑ 1 copy of the Fo and Missterssippi River Soil Run-off Procedure Sheet (master on page 75)

❑ 1 copy of the Rafta River Soil Run-off Procedure Sheet (master on page 76)

❑ 8 copies of the Sediment Test Procedure Sheet (master on page 106)

❑ 2 copies of the Phosphate Test Procedure Sheet (master on page 165)

- ❏ 4 copies of the Water Life Procedure Sheet (master on page 166)
- ❏ 16 copies of the Birdwatching Procedure Sheet (master on page 169)

Additional material for dice version of Population Game in Activity 5:

- ❏ 1 overhead transparency of the Random Attack Board (master on page 133)
- ❏ 320 dice (see note about dice on page 115)
- ❏ 16 small plastic bags for storing dice
- ❏ 1 spray bottle of water for "plumping" dice (only needed for sponge dice)
- ❏ 16 copies of the Random Attack Board (master on page 133)

Optional:

- ❏ live daphnia and three cups (see "Sources" on page 222 for more information)
- ❏ 12 name badges: Ken Unballe, Juan Tunó, Chairperson, Chemist, Mandy Lyfbotes, Avery Wun, Don Juan Tunó, LaToya Faktorie, Anton Alogue, Elmo Skeeto, Sandy Trapp, and Bo Vyne
- ❏ a shoebox in which to toss dice (only for dice version of Population Game in Activity 5)
- ❏ 8 sets of James Pond Clue Cards (masters on pages 173–176)

## Consumables

- ❏ 32 copies of the Map of the Gray Area (master on page 25)
- ❏ 32 copies of the Gray Area History Timeline (master on page 26)
- ❏ about 200 ml white vinegar (2 cups)
- ❏ several cups of distilled water
- ❏ 1–2 rolls of pH paper
- ❏ about 1 cup of soil
- ❏ about 2 tablespoons baking soda
- ❏ 2 coffee filter papers about 10" in diameter, flat or cone-shaped (paper towels may be substituted)
- ❏ 32 copies of the Test Results student data sheet (masters on pages 78–79)
- ❏ 32 copies of the Newspaper Interview (master on page 81)
- ❏ paper towels
- ❏ 4 or more tablespoons of soil (You need soil that will run off when "rained" on. Use any plain soil that isn't too rocky, clay-like, or too full of organic

material like roots and leaves. Avoid potting soil.)
- ❏ a handful of one or more of the following: pencil sharpener shavings, grass clippings, moss, dried weeds, or cut up pieces of paper towel
- ❏ 1 half-pint of chocolate milk
- ❏ 32 copies of the Sediments: What Do You Think? student sheet (master on page 111)
- ❏ a chalkboard, butcher paper, or overhead transparency copy of the Population Game Graph (master on page 132) and the Birdwatching graph (master on the Test Results data sheet, page 79)
- ❏ 1 small squeeze bottle of bromothymol blue (BTB)

Additional material for dice version of Population Game in Activity 5:
- ❏ 16 copies of the Population Game Graph (master on page 132)

Optional:
- ❏ for class work or homework: 32 copies of the
  - _ Chlorine: What Do You Think? worksheet (master on page 51)
  - _ Acid Rain: What Do You Think? worksheet (master on page 89)
  - _ Deer Lion: What Do You Think? worksheet (master on page 140)
  - _ James Pond: What Do You Think? worksheet (master on page 183)

## General Supplies

- ❏ scissors or a paper cutter
- ❏ a 3-hole punch
- ❏ an overhead projector
- ❏ 1 large black marker
- ❏ a permanent marker to write on masking tape
- ❏ 1 black permanent marker
- ❏ 8 large markers
- ❏ a meter stick or yardstick
- ❏ 1 piece of butcher paper for a large class map
- ❏ 1 piece of butcher paper about 8' x 4' for Suspect Chart (or equivalent chalkboard space)
- ❏ masking tape, or (for Activity 4 only) light-colored labeling tape (preferably white)
- ❏ paper clips
- ❏ 32 3-ring binders or 3-pronged folders for an Environmental Detective Notebook
- ❏ several sheets of lined paper for notes

- ❏ 40  9" x 12" envelopes or file folders
- ❏ a selection of crayons or colored pencils
- ❏ 2 packages large (2" square or larger) post-its
- ❏ 8 unsharpened pencils
- ❏ 2 colors of chalk, marker, or transparency pen
- ❏ 32 or more sheets of 8 ½" x 11" paper

Additional material for dice version of Population
Game in Activity 5:

- ❏ 16 sheets of scratch paper
- ❏ 16 pens each of two different colors, or 16 pens and
16 pencils

Mr. Harder
Jessica

## Mystery Conclusion

1. Which suspect(s) do you think
is /are most responsible for the fish
kill and why? Explain in details.

I think that Juan Tuno is killing
the fish. First of all, kids often lie when
they do something wrong. Juan
could've poisoned the fish with
something on accident, and lied
to everyone. Second of all, on
the chlorine tests, he could've added
chlorine to the samples before
he took them back to a lab.
This also goes for the sediment,
acid rain, and water life tests.
For the birdwatching tests, he
could have caged the birds
and let them out at the
area of the tests. In conclusion,
I think Juan Tuno did it
and was too afraid to tell everyone.

# Activity 1: Introducing the Mystery

## Overview

Over the weeks of the *Environmental Detectives* unit, students will become quite familiar with the complex environment of the fictitious, but realistic, "Gray Area." In this first activity, students are introduced to the mystery— why are fish dying in the waterways of the Gray Area?

Students are invited to play the roles of environmental scientists to solve the mystery through testing, research, discussion, and debate. They begin by brainstorming information about the needs of fish, and ideas about what might be causing the fish kill. They then use clue cards to create a timeline of the history of the area in their Environmental Detective Notebooks. They meet three characters from the area, and hear their statements.

*The timeline that students create in this first session will be used as a reference several times during the unit.*

The class generates a list of possible causes of the fish die-off, including pollutants like chlorine, acid rain, phosphates, sediment, and oil. They learn that, as environmental detectives, they will get to investigate these possibilities during the coming weeks.

*Note: Teachers of younger students may want to split Activity 1 into two sessions. The first session would include the introduction, the first brainstorming discussion, and starting their Environmental Detective Notebooks. The second session would include the timeline activity, introduction of suspects, and outline of the unit.*

## What You Need

### For the class:

- ❏ scissors or a paper cutter
- ❏ a 3-hole punch
- ❏ an overhead projector
- ❏ 1 overhead transparency of the Map of the Gray Area (master on page 25)
- ❏ 1 overhead transparency of the Gray Area History Timeline (master on page 26)
- ❏ 1 large black marker
- ❏ a meter stick or yardstick
- ❏ 1 piece of butcher paper for a large class map
- ❏ 1 piece of butcher paper about 8' x 4' for Suspect Chart (or equivalent chalkboard space)
- ❏ masking tape
- ❏ 1 copy of Juan Tunó's picture and Introductory statement (master on page 27)
- ❏ 1 copy of Don Juan Tunó's picture and Introductory statement (master on page 28)
- ❏ 1 copy of Avery Wun's picture and Introductory statement (master on page 29)

*You may choose to copy the suspect's pictures and statements a few at a time as they are needed for each activity, or all at one time. The "What You Need for the Whole Unit" on page 8 can help you if you choose the latter.*

**For each group of four students:**
- ❐ 1 set of Timeline Clue Cards (masters on pages 30–31)
- ❐ a paper clip

**For each student:**
- ❐ a 3-ring binder or 3-pronged folder for an Environmental Detective Notebook
- ❐ 1 copy of the Map of the Gray Area (master on page 25)
- ❐ 1 copy of the Gray Area History Timeline (master on page 26)
- ❐ several sheets of lined paper for notes

# Getting Ready

## Before the Day of the Activity

### Setting Up Headquarters

1. Prepare the Suspect Chart. This chart will be used by your students throughout the unit to record their ideas about who is guilty in the case of the dying fish. Write "Suspects" on the top, 8' edge of a piece of butcher paper or section of chalkboard. Make ten vertical columns 8 ½" wide, so the suspect pictures will fit at the tops of the columns. You will add the nine suspect pictures to the chart as the unit progresses. Label the tenth column with a big question mark. The tenth column is provided so that students record clues and ideas that don't necessarily implicate a specific character.

*The columns will later be used for students to vote, using post-its, on who they think is guilty. If you are teaching the unit to multiple classes, you may need to use different colored post-its to differentiate the classes' predictions on the chart.*

8 ½"    8 ½"

Suspect photos will go here as they are added each day.

| | | | | SUSPECTS | | | | | ? |
|---|---|---|---|---|---|---|---|---|---|
| | | | | | | | | | |

2. Make one copy each of the picture and Introductory statement for Juan Tunó (master on page 27), Don Juan Tunó (master on page 28), and Avery Wun (master on page 29).

*This is one example of what the Suspect Chart will look like towards the end of the unit:*

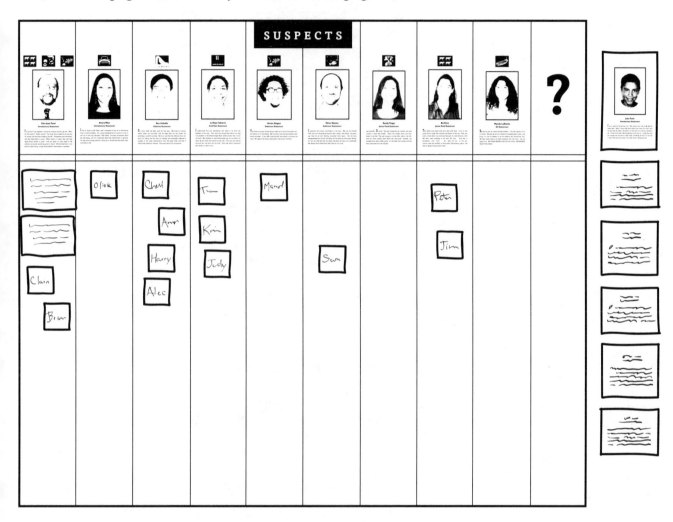

3. Make an overhead transparency of the Gray Area History Timeline (master on page 26).

4. Make the large class map.

    a. Use a photocopy machine to make an overhead transparency of the Map of the Gray Area (master on page 25).

    b. Project this transparency onto a piece of butcher paper of whatever size you would like your map to be. It should be large enough to be visible for the whole class.

    c. Use the large black marker to trace the projected map onto the butcher paper. Don't worry too much about details.

Juan Tunó is **not** a suspect, so his photo and statements will be nearby.

Juan Tunó's statements are added each day.

## Preparing for Teamwork

1. Group your students. On most days of the unit, students will work in groups of two, then later join with another group of two as a group of four for research and discussion. Since these groups will be working together so much, it's worth carefully selecting them prior to starting the unit. Try to make sure that each group has a balance of reading skill levels and leadership abilities.

2. Prepare the clue cards. Make enough copies of the Timeline Clue Cards (masters on pages 30–31) so you have one set of 14 for each group of four students. Cut up the cards, and clip each set with a paper clip. These can be collected and re-used with multiple classes.

3. Environmental Detective Notebooks. Each student will need to assemble and refer to quite a few papers and notes during the unit. Three-ring binders work well to prevent loss of loose pages. If you can't be sure to get binders or three-pronged folders for all students, plan to provide construction paper for folders.

4. Copy pages for the Environmental Detective Notebooks. Make a copy for each student of the Gray Area History Timeline (master on page 26) and the Map of the Gray Area (master on page 25). Students will put these in their Environmental Detective Notebooks. If their notebooks will be in binders, punch binder holes in the pages.

5. Read the Timeline Clue Cards yourself and familiarize yourself with the map and character statements for this session.

## On the Day of the Activity

1. Tape the large map and Suspect Chart to a wall. If possible, put them together in one area.

Headquarters

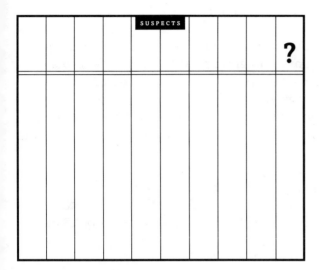

2. Have the rest of the materials handy for use after you have introduced the mystery.

## Introducing the Mystery of the Dying Fish

1. Announce that the class will get to study environmental science for the next few weeks. If needed, define the environment as the place or area surrounding living things. [air, water, land, weather, etc.]

2. Show students the large map of the Gray Area, and explain that it is a make-believe environment, but is based on real places.

3. Announce that the Gray Area has a problem. Fish have been dying off in large numbers. Here's some more information to share with students:

- About one thousand fish per year have died during each of the last five years.

- Most of the dead fish have washed up on the shores near Synchrony (SINK-krony) City.

*Note: Juan Tunó is a recurring character whose statements are read by the teacher as a means of introducing new ideas and information throughout the unit. The other characters are read by students. Plan to choose students who are comfortable reading in front of the class. To enliven the proceedings, you may plan to have them represent that character for the remainder of that activity, wearing a name badge, simple costume, or hat, depending on whether or not you think your students would enjoy doing this.*

*You might want to point out some of the rivers and lakes and their pronunciations, especially Lake Adaysicle (a-DAY-zickle; as in the word "lackadaisical") and the Fo (rhymes with "oh") River.*

- The people who live in the Gray Area are bothered by the bad smell, and worried that all the dead fish might mean that something is wrong in their environment.

- Autopsies have been ordered on samples of dead fish from all bodies of water, but the results will not be back for at least a month.

4. Tell the class that, in the next few weeks, they will get to try to solve the **environmental mystery** of what is responsible for killing the fish in the Gray Area. Like real environmental scientists, they will conduct tests, do research to identify possible environmental problems, causes, and issues, and then attempt to solve the mystery.

5. Ask what a detective does. [Searches for clues and information and puts them together to solve mysteries—very much like what scientists do.] Tell the students that they'll not only be scientists, they'll also be "environmental detectives."

6. Use the following questions to lead a brief discussion:

- What are the needs of fish? [food, water, shelter, oxygen, space, clean water, etc.]

- Looking at the map, what kinds of factors in the Gray Area environment might affect their needs?

- How might a large fish kill affect other living things and the rest of the environment?

- How might the students find information to help figure out what's going on in the area?

- What do the students think might be killing the fish?

## Creating a Timeline of the History of the Gray Area

1. After accepting their ideas, tell the students that in the coming days they will get to do some of the investigations they've suggested. Tell the students it may help them to start by looking for clues in the **history** of the area.

2. Distribute copies of the map, the timeline, and lined paper. Ask everyone to put them in their binders or folders to start an "Environmental Detective Notebook." Tell students that they are responsible for bringing the notebook to each lesson of the unit. They'll record test results, ideas, and their research in this notebook.

3. On the overhead transparency of the Gray Area History Timeline, model how to use the clue cards to fill in the timeline. Read a clue card as an example: *"The cattle ranch was started 75 years ago."* Emphasize that they don't have to record every word—notes and abbreviations are fine. For instance, show how to write "cattle ranch started" with a line pointing to 75 years ago.

4. Make sure students understand that the last 20 years of the top timeline are "spread out" (expanded) on the bottom timeline to give them more room to write.

5. All students should record every event on their own timelines, but they will share the clue cards with their group of four. Have one student from each group get a set of clue cards, and have them begin.

6. Circulate, encouraging students to discuss within their group any discoveries they make or theories they may have about the dying fish.

7. When students have finished recording, have them reassemble and clip together the clue cards and return them to the front of the classroom.

## Timeline Discussion

1. Regain the attention of the whole class. Ask your students if they found any clues that might explain why the fish are dying. Allow them to share and discuss their ideas. [Students may point out that the water slide opened at the same time as the fish started dying.]

2. After accepting all their ideas, tell them that the Board of Supervisors of the Gray Area (similar to a City Council) suspects that the **water slide** is the culprit, and they plan to close it down. Let them know that many people in the area are upset about this, and a local student, Juan Tunó, has started his own investigation.

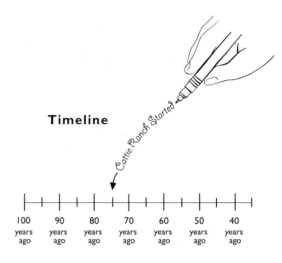

**Timeline**

Cattle Ranch Started

| | | | | | | |
|100 years ago|90 years ago|80 years ago|70 years ago|60 years ago|50 years ago|40 years ago|

*If someone mentions chlorine, ask students what they know or have heard about chlorine.*

3. Show the picture of Juan Tunó, read his statement, then tape it on the wall *beside* the Suspect Chart. Emphasize that he is **not** a suspect.

## Juan Tunó
### Introductory Statement

I go to school in Synchrony City. I like science, but I also really like the water slide. When I heard that they might close down the water slide because fish are dying, I decided to do my own tests with my chemistry kit. I found chemicals called phosphates in the stream. I think they may be coming from my uncle's cattle ranch, and killing the fish. It's not fair to shut down the water slide if the cattle ranch is killing the fish.

4. Tell your students that Juan's uncle, Don Juan Tunó, has a different opinion. Show your students the picture of Don Juan Tunó, ask a student volunteer to read his statement, then tape it on the Suspect Chart.

## Don Juan Tunó
### Introductory Statement

I'm proud of my nephew's interest in science, but let's get real. What do fish matter? People matter! The Gray Area should not be wasting its money and resources studying the fish. Phosphates are in fertilizers and they help plants to grow. When they're in water, they also help algae (water plants) to grow. Fish eat algae, so more algae in the streams and ponds would be good for them! Tell the fishermen to cut back on their fishing to stop the problem—if there even *is* a problem.

5. Ask your students if they have any comments or observations on Don Juan Tunó's statement. [They will probably mention that Don Juan Tunó is the owner of Tunó Enterprises, which owns the cattle ranch, the logging company, and the oil refinery. **Don't tell the students now,** but later on he will probably be found "not guilty" by your students, based on their test results.]

6. Tell students that Juan Tunó has persuaded the Gray Area Board of Supervisors not to shut down the water slide yet. However, the Board has specified that they need to make a decision about the water slide in ____ *(however long your unit will last).*

## Outlining the Unit for the Students

1. Ask your students what they think could be getting into the water and killing the fish. Accept all their ideas and list them on the chalkboard.

2. Say that they will get to do research and conduct scientific tests to gather information about some of these possible problems. Make sure the following substances are part of the list:

- chlorine
- acid rain
- dirt (or sediment)
- phosphates
- oil

3. Put a star next to each of the five substances that students will investigate during the unit. Explain that they will find out if any of these substances are getting into the water and, if so, what's causing that to happen.

4. Tell students the area where the chart and the map are located will be called "Headquarters." Explain that during the *Environmental Detectives* unit, they will investigate these possible problems, and will meet the people suspected of causing each problem. Each "suspect's" statements will be added to the chart.

5. Ask students to be patient, because it may take time to make a final decision as to who they think is guilty.

*We have chosen to use the term "suspect" even though it is not a term normally used in solving environmental problems. The true suspects are often not individuals, but chemical causes, environmental factors, institutional policies or practices, etc. We've chosen to put human faces on these problems to make them more interesting to students.*

6. Tell the class there is another suspect for them to meet now. Show them the picture of Avery Wun, ask a student to read the statement, then tape it on the Suspect Chart.

## Avery Wun
### Introductory Statement

I live in Gray's Land Town, and I commute in my car to Synchrony City to work everyday. It's a long commute, but it's worth it to me to live out of the city, because I like nature. I'm very concerned about the fish dying, and I'm concerned about the environment in general. I'm just your average person—I shop, go to movies and the water slide, and I like to fish.

7. Give the class a few minutes to write in their detective notebooks about anything they have learned so far that might explain why the fish are dying. They should also make a list of questions. What do they want to find out more about to help solve the mystery?

# Map of the Gray Area

Map labels: Selectively Cut Forest, Upper Rafta River, Clear Cut Forest, Lake Adaysicle, Dam, Lower Rafta River, Parallel Park, Farm, Synchrony City, Gray Bay, Miss., Lower, Upper Missterssippi, Toy Factory, Cattle Ranch, Fo River, Oil Refinery, Gray's Land Town, James Pond, Water Slide, Golf Course

# Gray Area History Timeline

**Events *before* 100 years ago:** 300 years ago—Gray Area Discovered
150 years ago—Synchrony City Built

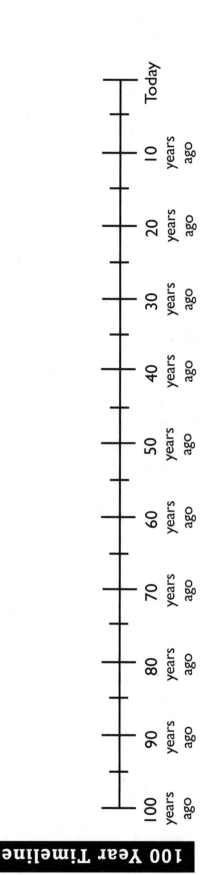

**100 Year Timeline**

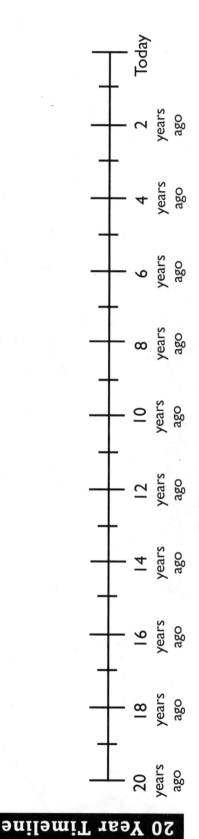

**20 Year Timeline**

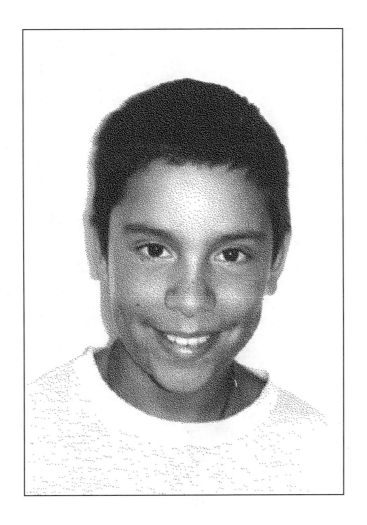

## Juan Tunó
## Introductory Statement

I go to school in Synchrony City. I like science, but I also really like the water slide. When I heard that they might close down the water slide because fish are dying, I decided to do my own tests with my chemistry kit. I found chemicals called phosphates in the stream. I think they may be coming from my uncle's cattle ranch, and killing the fish. It's not fair to shut down the water slide if the cattle ranch is killing the fish.

## Don Juan Tunó

### Introductory Statement

I'm proud of my nephew's interest in science, but let's get real. What do fish matter? People matter! The Gray Area should not be wasting its money and resources studying the fish. Phosphates are in fertilizers and they help plants to grow. When they're in water, they also help algae (water plants) to grow. Fish eat algae, so more algae in the streams and ponds would be good for them! Tell the fishermen to cut back on their fishing to stop the problem—if there even *is* a problem.

## Avery Wun

### Introductory Statement

I live in Gray's Land Town, and I commute in my car to Synchrony City to work everyday. It's a long commute, but it's worth it to me to live out of the city, because I like nature. I'm very concerned about the fish dying, and I'm concerned about the environment in general. I'm just your average person—I shop, go to movies and the water slide, and I like to fish.

The area was discovered 300 years ago by Earl Gray, and was named after him.

The fish started dying off 5 years ago. Most of the dead fish have been found floating in the water or washed up on the shore near Synchrony City.

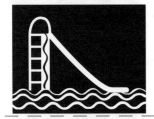

The water slide opened 5 years ago. Owned by Ken Unballe.

Parallel Park was founded 45 years ago. Deer, duck, and rabbit hunting are allowed there, but very restricted. Fishing is very popular in the whole Gray Area.

The hunting of mountain lions in Parallel Park was stopped 20 years ago, then started up again 6 years ago.

The cattle ranch was started 75 years ago.

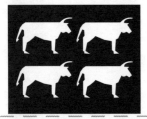

The cattle ranch doubled the number of cattle 7 years ago. Owned by Tunó Enterprises.

The golf course opened 8 years ago. Owned by Sandy Trapp.

Gray's Land Town was built 50 years ago where there used to be parsley farms. It has grown a lot in the last 10 years, with many people living in the small town, and commuting to Synchrony City to work.

**TIMELINE CLUE CARD • TIMELINE CLUE CARD • TIMELINE CLUE CARD • TIMELINE CLUE CARD**

Synchrony City was built 150 years ago. The city has seen a lot of growth in the last 10 years, with many houses in the suburbs being built, and freeways built for the many commuters.

**TIMELINE CLUE CARD • TIMELINE CLUE CARD • TIMELINE CLUE CARD • TIMELINE CLUE CARD**

The oil refinery started operating 18 years ago. Owned by Tunó Enterprises.

**TIMELINE CLUE CARD • TIMELINE CLUE CARD • TIMELINE CLUE CARD • TIMELINE CLUE CARD**

Logging in the area began 100 years ago. There are no old growth forests left in the area.

**TIMELINE CLUE CARD • TIMELINE CLUE CARD • TIMELINE CLUE CARD • TIMELINE CLUE CARD**

To prevent erosion, the logging companies started selective cut logging in the area 12 years ago. They still do some clear cut logging too. Owned by Tunó Enterprises.

**TIMELINE CLUE CARD • TIMELINE CLUE CARD • TIMELINE CLUE CARD • TIMELINE CLUE CARD**

The Toy Factory opened 17 years ago. Owned by LaToya Faktorie.

**TIMELINE CLUE CARD • TIMELINE CLUE CARD • TIMELINE CLUE CARD • TIMELINE CLUE CARD**

# Activity 2: Chlorine Tests

## Overview

In this activity, students "attend" a meeting of the Gray Area Board of Supervisors, and learn more about possible chlorine pollution in the area, related to the water slide. They find out the results of a chemical test for chlorine that was done earlier.

Next, teams of students analyze, interpret, and discuss the contents of the "Chlorine Files," submitted to the Board. These files include data from both chemical and biological tests for chlorine, related newspaper articles, a water slide schedule, reference materials, and even some "secret documents." Through their research, students learn about the use of bioindicators and the value of doing both chemical and biological tests. During small and large group discussions, they attempt to resolve the seemingly contradictory results of the two tests.

At the end of the activity, they are given an opportunity to reflect and write notes in their Environmental Detective Notebooks. An optional worksheet can be used in class or as homework for students to record their results and any ideas about chlorine.

*Note: The biological test data students analyze in this activity only involve paper and pencil. By looking at pictures, students compare the numbers of small, aquatic crustaceans called daphnia (or water fleas) found in different parts of the Fo River. **If possible, we strongly recommend that you use real daphnia instead of the paper and pencil version on page 45.** Information on how to obtain and set up samples of daphnia is in the "Sources" section on page 222.*

## What You Need

**For the class:**

- ❐ Headquarters set-up from previous session
- ❐ 4 copies of the record of the Meeting of the Gray Area Board of Supervisors (master on page 43–44)
- ❐ masking tape
- ❐ 1 copy of the Water Flea Diagram (master on page 42)
- ❐ 1 copy of Juan Tunó's Chlorine statement (master on page 40)
- ❐ 1 copy of Ken Unballe's picture and Chlorine statement (master on page 41)
- ❐ *(optional)* live daphnia and three cups (see "Sources" on page 222 for more information)
- ❐ *(optional)* 4 name badges: Ken Unballe, Juan Tunó, Chairperson, and Chemist

*You may want to show a daphnia video to your students. See the "Resources" section on page 230.*

### For each group of four students:
- ❏ a 9" x 12" envelope or file folder
- ❏ 1 copy of the Chlorine Files (masters on pages 45–49)
- ❏ 1 Chlorine Discussion Card (master on page 50)

### For each student:
- ❏ their Environmental Detective Notebooks from previous session
- ❏ *(optional)* for class work or homework: 1 copy of the Chlorine: What Do You Think? worksheet (master on page 51)

## Getting Ready

### Before the Day of the Activity

1. Remind students to bring their Environmental Detective Notebooks to this and every lesson in the unit.

2. Make four copies of the record of the Meeting of the Gray Area Board of Supervisors (master on pages 43–44). Also make a copy of the Water Flea Diagram (master on page 42). If you'd like to use them, make name badges to help identify the four speakers in the meeting.

3. Prepare the Chlorine Files (masters on pages 45–49). Make enough copies of the articles and documents so there is one set per group of four. Put each set in an envelope. Label each envelope "Chlorine Files."

4. Copy and cut apart the Chlorine Discussion Cards (master on page 50) so there is one card per group of four.

5. Copy Juan Tunó's Chlorine statement (master on page 40) and tape it up at Headquarters under his picture and first statement.

*The record of the board meeting will be read aloud by four students during the class session. You may want to choose four students ahead of time, and give them a few minutes before class to read over their parts. Be sure the student reading the part of Juan Tunó knows how to pronounce **bioindicator** (by-oh-IN-dih-kay-tor) and **daphnia** (DAF-nee-ah).*

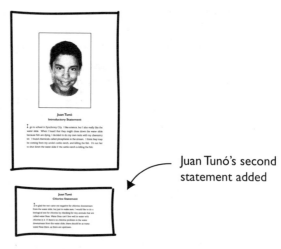

Juan Tunó's second statement added

6. Copy Ken Unballe's picture and Chlorine statement (master on page 41) and tape it on the Suspect Chart at Headquarters.

| | | | | SUSPECTS | | | | | | ? |
|---|---|---|---|---|---|---|---|---|---|---|
| | | | | | | | | | | |

7. If you've decided to use them, make one copy of the Chlorine: What Do You Think? worksheet (master on page 51) for each student.

## Introducing the Chlorine Problem

1. Announce that the Gray Area Board of Supervisors met recently about whether the water slide has something to do with the fish dying. Say you have copies of the record of the meeting. Ask for four volunteer students to read the parts of the speakers.

2. Have the four students come to the front of the class, give them each a copy of the meeting record (and name badges, if you made them), and have them begin. Provide the student playing the role of Juan Tunó with a copy of the Water Flea Diagram and an envelope containing the Chlorine Files.

3. Point out to the class that statements for Ken Unballe and Juan Tunó are on the wall at Headquarters.

*If you have decided to use live daphnia, set up and label three containers—one to represent a healthy stream, one for the area of the Fo River upstream from the water slide, and the third for downstream from the water slide. Both the healthy stream and upstream containers should contain many daphnia, while the downstream container should contain very few—similar to the illustrations for the results of the biological testing on page 1 of the Chlorine Files. Set the containers out for students to use when they review the Chlorine Files.*

*You may want to clarify the meaning of "upstream" and "downstream" from the water slide, by pointing out on the map that the Fo River flows **downhill** past the water slide toward Gray Bay.*

# MEETING OF THE GRAY AREA BOARD OF SUPERVISORS

**Four Speakers:** The Chairperson, Ken Unballe,
A Chemist, and Juan Tunó

**Chairperson:** The whole Gray Area is talking about the mystery of the dying fish. Everywhere I go, people tell me what they think is causing the fish to die. It's getting out of control, and it's time we got to the bottom of this. Some people think it's suspicious that the fish began dying around the time the water slide opened five years ago. Let's begin by hearing from the owner of the water slide.

**Ken Unballe:** Members of the Board, our water slide has been great for this area. We bring in tourists, which helps the economy, and we keep kids off the streets. We put chlorine in the water to kill germs so our water is safe for kids. We're sure that the chlorine from our water slide isn't killing the fish, and here's why: We collected a sample of the water in the river downstream from the water slide, and we had it chemically tested for chlorine. Our chemist is here with the results.

**Chemist:** As most of you know, used water is released from pipes at the water slide, then flows downstream in the Fo River, toward Gray Bay. I tested the water downstream from the water slide. I did the test on Tuesday at 3:00 p.m. The results were negative. There was NO CHLORINE in the river downstream from the water slide.

**Ken Unballe:** Excuse me but could you please repeat that? How much chlorine did you say was in the Fo River?

**Chemist:** None! Zilch! Nada! Goose egg!

**Ken Unballe:** Thank you. As the Board can see, we're a responsible business and we care about the environment. We wish you good luck in finding out what is killing the fish, but we've proven clearly that it's not us.

• 1 •

**Juan Tunó:** I've been going to the water slide since it opened, and I'm glad the test came out negative for chlorine. But just to make sure, I decided to do a biological test for chlorine by checking for tiny animals that are called water fleas, also known as daphnia.

**Ken Unballe:** What?! We've already heard from a chemist. We don't need any more proof—especially not from a kid with fleas!

**Juan Tunó:** Water fleas, sir. They're different from regular fleas. The water fleas might tell us something the chemist can't. For example, coal miners used to take canaries into the mines. If the sensitive canary suddenly died, the miners knew that there must be poisonous gas in the mine, and they'd better get out. The canary was a **bioindicator—** a living thing used to test for something that was hard for humans to detect.

**Ken Unballe:** Birds and fleas are beside the point.

**Juan Tunó:** Anyway, I used water fleas as bioindicators to test for chlorine. *(Holds up the diagram of a water flea.)* Water fleas can't live well in water with chlorine in it. If there's no chlorine problem downstream from the water slide, there should be just as many water fleas downstream as there are upstream. My results are right in this file. I've also been clipping some newspaper articles about chlorine that I've included. Also, some other documents were left on my front porch in an envelope…

**Ken Unballe:** *(Getting angry and interrupting Juan Tunó.)* I really must object! The Board knows how much good the water slide does. This is unnecessary! This isn't a Watergate or Whitewater scandal! This is a harmless water slide!

**Chairperson:** Order! Order! The Board needs time to review the evidence and make a responsible decision. Everyone, please just leave your test results and other information, and we'll get back to you. Thank you. Meeting adjourned.

• 2 •

## Small Group Study and Discussion

1. Tell students they will now receive copies of the documents the Board received, and they will try to decide whether chlorine from the water slide is causing the fish to die.

2. In their group's copy of the Chlorine File they'll find:

- The results of the chemical and biological tests on chlorine
- Juan Tunó's newspaper clippings
- E-mails, notes, and documents from an envelope left near Juan Tunó's front door
- Reference material about water fleas and bioindicators

3. When each group receives the file, they will read the documents and search for clues. Any of the clues could be important. **Caution students not to write on the documents in their files.**

4. Tell your students that when the team has finished reading and talking over all the documents in the file, they will send one person to get a Chlorine Discussion Card. This card has questions to help them further discuss their ideas.

5. Distribute the Chlorine Files and have teams begin.

## Large Group Discussion

1. Refocus the entire class, and ask one group to summarize the results of Juan Tunó's chlorine **biological** test. Ask the other groups if they agree. [There are far fewer water fleas downstream from the water slide than upstream.]

2. Ask, "What does the lack of water fleas downstream from the water slide tell us?" [Something, possibly chlorine, is killing the water fleas.]

3. Ask, "What were the results from the **chemical** test for chlorine?" [No chlorine was found.]

4. Ask, "Why might the results of the two chlorine tests be different?" Depending on their research discussion of

*While the research and communication skills developed in this and future sessions are extremely valuable, they do take practice. If your students have had little previous research experience, spend some time as a class going over some good ways to approach the files. You might suggest these ideas:*

- *The group can quickly look over and "inventory" the contents of the file together to see how many documents there are and which ones seem to be most important.*

- *The group can divide up the documents. Each student in the small group can be responsible for reading at least two documents.*

- *As they read, each student can take notes about things they read that might help solve the mystery.*

- *Each student can tell the rest of the group what they read.*

*Note: The discussion questions here are intended to help you stimulate debate. Of course the direction of a discussion can't be entirely anticipated, so please use these only as a general guide. As much as possible, the ideas and information should come from the students. Likely student responses have been included in brackets, but are not meant to be read to the students or used to correct their answers. If your students bring up a particular chart or article during the discussion, direct the attention of the entire class to their own copies of the chart or article and discuss it together. There are many valid directions the discussion may go.*

the Chlorine Files, students may bring up the following distinction between chemical and biological tests:

> **Chemical tests** are accurate, but they test the water at only the particular moment when the sample is collected. Even though the Tuesday test was negative, it doesn't mean tests on samples from other days and times would be negative.

> **Biological tests** are more accurate for detecting longterm impact. A test can show, for example, that an organism that is sensitive to a particular chemical is not found in a location where it would usually be. If so, there is a high probability that the chemical is sometimes there, even if it's not present at the moment.

5. Tell the class that environmental scientists often use **both** biological and chemical tests. If your students do not bring it up, ask them to look at the weekly water slide schedule to see when the sample water was collected for the chemical test. [It was collected almost a full week after the last release of used water into the river.]

6. Students may believe they have already solved the mystery of the dying fish. Point out that the mystery may be complicated, and encourage them not to jump to conclusions too quickly.

## Reflecting on Chlorine

1. Tell the students they will have some time now to:

- Write about whether they think the chlorine is killing the fish. Why and how? Which test results should we believe?

- Write notes about suspects in their detective notebooks, and under the suspects' pictures.

2. Have students reassemble their team's Chlorine Files. Tell students the file will be at Headquarters and can be borrowed if they need to look at it again. Collect the files.

3. If you've decided to use it, assign as homework (or have students work in class on) the Chlorine: What Do You Think? worksheet. Have students put this sheet in their Environmental Detective Notebooks.

# Juan Tunó
## Chlorine Statement

I'm glad the test came out negative for chlorine downstream from the water slide, but just to make sure, I would like to do a biological test for chlorine by checking for tiny animals that are called water fleas. Water fleas can't live well in water with chlorine in it. If there's no chlorine problem in the water downstream from the water slide, there should be as many water fleas there, as there are upstream.

## Ken Unballe
### Chlorine Statement

Our water slide has been great for this area. We bring in tourists, which helps the economy, and we keep kids off the streets by providing a positive activity. We're so sure that the chlorine from our park isn't killing the fish, that on Tuesday we voluntarily collected a sample of the water downstream from the water slide and had it chemically tested for chlorine. The results prove our innocence!

**Water Flea**

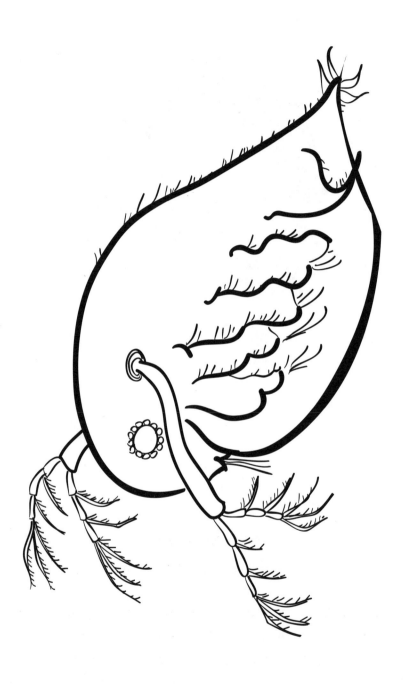

# MEETING OF THE GRAY AREA BOARD OF SUPERVISORS

**Four Speakers:** The Chairperson, Ken Unballe,
A Chemist, and Juan Tunó

**Chairperson:** The whole Gray Area is talking about the mystery of the dying fish. Everywhere I go, people tell me what they think is causing the fish to die. It's getting out of control, and it's time we got to the bottom of this. Some people think it's suspicious that the fish began dying around the time the water slide opened five years ago. Let's begin by hearing from the owner of the water slide.

**Ken Unballe:** Members of the Board, our water slide has been great for this area. We bring in tourists, which helps the economy, and we keep kids off the streets. We put chlorine in the water to kill germs so our water is safe for kids. We're sure that the chlorine from our water slide isn't killing the fish, and here's why: We collected a sample of the water in the river downstream from the water slide, and we had it chemically tested for chlorine. Our chemist is here with the results.

**Chemist:** As most of you know, used water is released from pipes at the water slide, then flows downstream in the Fo River, toward Gray Bay. I tested the water downstream from the water slide. I did the test on Tuesday at 3:00 p.m. The results were negative. There was NO CHLORINE in the river downstream from the water slide.

**Ken Unballe:** Excuse me but could you please repeat that? How much chlorine did you say was in the Fo River?

**Chemist:** None! Zilch! Nada! Goose egg!

**Ken Unballe:** Thank you. As the Board can see, we're a responsible business and we care about the environment. We wish you good luck in finding out what is killing the fish, but we've proven clearly that it's not us.

• 1 •

# MEETING OF THE GRAY AREA BOARD OF SUPERVISORS
## —continued

**Juan Tunó:** I've been going to the water slide since it opened, and I'm glad the test came out negative for chlorine. But just to make sure, I decided to do a biological test for chlorine by checking for tiny animals that are called water fleas, also known as daphnia.

**Ken Unballe:** What?! We've already heard from a chemist. We don't need any more proof—especially not from a kid with fleas!

**Juan Tunó:** Water fleas, sir. They're different from regular fleas. The water fleas might tell us something the chemist can't. For example, coal miners used to take canaries into the mines. If the sensitive canary suddenly died, the miners knew that there must be poisonous gas in the mine, and they'd better get out. The canary was a **bioindicator**— a living thing used to test for something that was hard for humans to detect.

**Ken Unballe:** Birds and fleas are beside the point.

**Juan Tunó:** Anyway, I used water fleas as bioindicators to test for chlorine. *(Holds up the diagram of a water flea.)* Water fleas can't live well in water with chlorine in it. If there's no chlorine problem downstream from the water slide, there should be just as many water fleas downstream as there are upstream. My results are right in this file. I've also been clipping some newspaper articles about chlorine that I've included. Also, some other documents were left on my front porch in an envelope…

**Ken Unballe:** *(Getting angry and interrupting Juan Tunó.)* I really must object! The Board knows how much good the water slide does. This is unnecessary! This isn't a Watergate or Whitewater scandal! This is a harmless water slide!

**Chairperson:** Order! Order! The Board needs time to review the evidence and make a responsible decision. Everyone, please just leave your test results and other information, and we'll get back to you. Thank you. Meeting adjourned.

• 2 •

## 1. Chemical Test Results for Chlorine:
Negative.  Percent chlorine: 0%

## 2. Biological Test Results for Chlorine:
The three samples of water below were gathered by Juan Tunó.  Please count the number of water fleas in the samples and analyze these results for yourself.

**Healthy Stream**

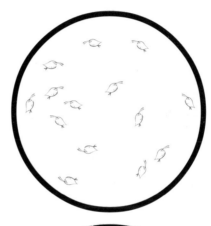

**Upstream from Water Slide**

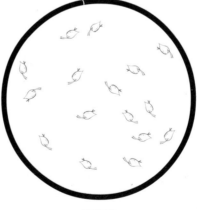

**Downstream from Water Slide**

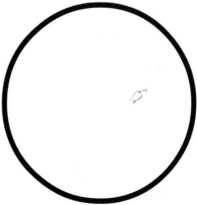

# Chlorine Spill

## Hundreds of fish and wildlife dead

*By Carolyn Tong*
BENCH PRESS

At Waving Waters water slide park, located 800 miles north of Gray Bay, workers accidentally spilled 250 gallons of chlorine yesterday, killing hundreds of fish and many other wildlife in Fiasco Creek.

When used safely, chlorine cleans our water, and prevents diseases. The World Health Organization estimates that 25,000 children die from waterborne diseases each day in parts of the world where chlorinated water is not used.

A6  *Lawrence Hall Gazette*  ☆☆☆☆☆    FRIDAY, JANUARY 19, 2001

# E. coli Outbreak

## E. coli from Water Park in Neighboring State Kills One Child

*By Jacquey Cort*
FULL COURT PRESS

One of the twenty-six children who got sick with E. coli bacteria in Shady Park wading pool died yesterday. She is the only child in the nation known to have died this way.

The children were probably contaminated by an infected child who was in the pool, scientists said. Normally bacteria are killed by chlorine in the pool, but in this case, bacteria survived long enough to spread to other children. Investigation continues into the pool's chlorine content.

SCIENCE

# Fo River Passes Chlorine Test

## Chemist tests river water and announces it's chlorine free

*By Felicia Barakos*
THE OLIVE PRESS

A local water slide park voluntarily collected samples from the Fo River downstream from where they release their used water. Chemists tested the river water and announced that there is no chlorine in it, and that it is safe for wildlife.

"I knew we weren't killing the fish," said Ken Unballe, owner of the park. "That's why I volunteered to collect the samples to be tested."

Mime-Version: 1.0
Date: Fri, 16 Oct 14:03:55
To: gungadin@slide
From: kenunball@slide (Ken Unballe)
Subject: Chlorine Tests

The Board of Supervisors is going to have a meeting about whether we should be shut down. We've got to keep the public on our side, and make this chlorine thing go away. Before the board meeting, I'm ordering the water downstream from our dump pipes to be tested for chlorine by a chemist, but check out the Water Slide Schedule, and I think you'll see why I think the chemist should **only** collect samples on **Tuesday night!**
Ken

---

Mime-Version: 1.0
Date: Thurs, 22 Oct 10:21:01
To: gungadin@slide
From: kenunball@slide (Ken Unballe)
Subject: more on chlorine tests

I thought we were finished with this after the chlorine test the chemist did—guess not. Now I hear that some mixed up kid thinks he is going to save the water slide by doing a different kind of test. He's going to test the water in the Fo River where we dump our water slide water, by counting water fleas. I guess the idea is that if there are water fleas, then there's not too much chlorine, because they're real sensitive to it, or something. Any ideas about what to do? This is getting serious!
Ken

---

Mime-Version: 1.0
Date: Fri, 23 Oct 08:22:07
To: gungadin@slide
From: kenunball@slide (Ken Unballe)
Subject: re: re: more on chlorine tests

I like your idea of buying a bunch of water fleas and dumping them in the water just before the kid does the test. Can you be in charge of ordering them and taking care of what needs to happen? Of course this should all be kept confidential.
Ken
P.S.—let's talk about a raise for you soon!

# Weekly Water Slide Schedule

| Monday | Tuesday | Wednesday | Thursday | Friday | Saturday | Sunday |
|---|---|---|---|---|---|---|
| 8:00am - 7:00pm Open to public | 8:00am - 7:00pm Open to public | Closed to public today | 8:00am - 7:00pm Open to public | 8:00am - 7:00pm Open to public | 8:00am - 7:00pm Open to public | 8:00am - 7:00pm Open to public |
| 7:00pm Bacteria tests | 7:00pm Bacteria tests | 6:00am Release used water into river | 7:00pm Bacteria tests | 7:00pm Bacteria tests | 7:00pm Bacteria tests | 7:00pm Bacteria tests |
| | 8:00pm Pump new water from river into holding tanks, and add chlorine | 10:00am Pump water from holding tank into pools | | | | |
| | | 1:00pm Bacteria tests | | | | |

# Biological Testing in the Environment

### What are Bioindicators?

Bioindicators (by-oh-IN-dih-kay-tors) are animals and plants that are sensitive to changes in their environment—they *indicate* or show that changes are taking place. Different bioindicators are sensitive to different types of changes. Scientists study changes in the populations of animal and plant bioindicators to see if an environment is healthy.

### Using Water Fleas as Bioindicators

Water fleas, also called daphnia, are tiny animals in the same family as shrimp. There are water fleas in almost every body of water. They eat algae and microscopic animals by sweeping them into their mouths with their waving legs. Many kinds of fish eat water fleas.

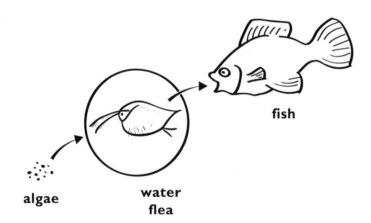

algae      water flea      fish

Because water fleas are very sensitive, we can use them as bioindicators to tell if there is a problem in the water. Small amounts of chlorine can kill water fleas, so a drop in their numbers tells us there may be chlorine pollution. A lack of water fleas could also mean:

- Not enough dissolved oxygen in the water for them to breathe.
- Water that is too acidic or too basic.
- Unhealthy amounts of a chemical other than chlorine.

## Chlorine Discussion Card

- What were the results of the chemical test for chlorine?
- What were the results of the biological test for chlorine using water fleas?
- Did the two tests agree?  Can you explain this result?
- Do you think the chlorine is killing the fish?  Why and how?

---

## Chlorine Discussion Card

- What were the results of the chemical test for chlorine?
- What were the results of the biological test for chlorine using water fleas?
- Did the two tests agree?  Can you explain this result?
- Do you think the chlorine is killing the fish?  Why and how?

---

## Chlorine Discussion Card

- What were the results of the chemical test for chlorine?
- What were the results of the biological test for chlorine using water fleas?
- Did the two tests agree?  Can you explain this result?
- Do you think the chlorine is killing the fish?  Why and how?

---

## Chlorine Discussion Card

- What were the results of the chemical test for chlorine?
- What were the results of the biological test for chlorine using water fleas?
- Did the two tests agree?  Can you explain this result?
- Do you think the chlorine is killing the fish?  Why and how?

# Chlorine: What Do You Think?

1. What did the **chemical test** for chlorine show? _____

_____

2. In the **biological test,** what did you find out about water fleas downstream from the water slide?

_____

_____

_____

3. Can you think of an explanation for the conflicting results?

_____

_____

_____

_____

_____

_____

4. Do you think chlorine is killing the water fleas?  Please explain.

_____

_____

5. Do you think the chlorine is killing the fish?  Please explain.

_____

_____

_____

# Activity 3: Acid Rain

## Overview

Although the last activity may indicate to your environmental scientists that chlorine from the water slide is the problem, they should not judge too soon. In this and subsequent activities, your class will explore several other possible reasons for the fish die-off.

The pH tests, research, and discussions conducted by students in this activity will take two class sessions. In the first session, students are introduced to the acid rain problem by Juan Tunó. Pairs of students circulate to 10 stations where they conduct pH tests on water samples from the Gray Area. They find that, although acid rain falls on the whole Gray Area, some rivers and lakes are more affected than others. At the end of the session or for homework, students read a newspaper "interview" with Juan Tunó in which he explains how acid rain is formed.

In the second session, student groups do research in the "Acid Rain Files" to answer some of their questions, including why some waterways are more acidic than others. They share their findings and ideas in a teacher-led class discussion. Gray Area characters introduce questions about the source of the air pollution that is thought to have caused the acid rain. At the end of the session, students are given time to reflect, write notes, and post their predictions on the Suspect Chart.

*Background information on acid rain is provided on pages 213–216.*

*If your students are not familiar with it already, we strongly suggest pre-teaching the concept of pH chemistry. The GEMS guide* Of Cabbages and Chemistry *is recommended for this purpose. The GEMS guide* Acid Rain, *or portions thereof, could also be used as a pre-teaching aid, or as a substantial and more in-depth extension. See the "Going Further" for this activity on page 70 for more about the* Acid Rain *unit.*

## What You Need

**For the class:**
- ❐ Headquarters set-up from previous sessions
- ❐ 9 plastic cups (8–10 oz. size)
- ❐ about 200 ml white vinegar (2 cups)
- ❐ several cups of distilled water
- ❐ masking tape
- ❐ a permanent marker to write on masking tape
- ❐ 1–2 rolls of pH paper (see "Getting Ready" below for more information)
- ❐ a cottage cheese type or other container (about 2 cup size)
- ❐ 8 copies of the pH Procedure Sheet (master on page 74)

- ❏ 1 copy of the Fo and Missterssippi River Soil Run-off Procedure Sheet (master on page 75)
- ❏ 1 copy of the Rafta River Soil Run-off Procedure Sheet (master on page 76)
- ❏ 3 copies of the pH Scale (master on page 77)
- ❏ a selection of crayons or colored pencils
- ❏ about 1 cup of soil
- ❏ about 2 tablespoons baking soda
- ❏ 2 coffee filter papers about 10" in diameter, flat or cone-shaped (paper towels may be substituted)
- ❏ 2 two-liter clear plastic soda bottles
- ❏ 1 pair of sharp scissors
- ❏ 1 copy of LaToya Faktorie's picture and Acid Rain statement (master on page 72)
- ❏ 1 copy of Juan Tunó's Acid Rain statement (master on page 71)
- ❏ 1 copy of Don Juan Tunó's Acid Rain statement (master on page 73)
- ❏ a chalkboard, butcher paper, or overhead transparency copy of the Acid Rain Questions (master on page 80)
- ❏ 1 package large (2" square or larger) post-its
- ❏ several small paper bags or garbage cans
- ❏ 1 or 2 sponges for cleaning up spills

## For each group of four students:
- ❏ a 9" x 12" envelope or file folder
- ❏ 1 copy of the Acid Rain Files (masters on pages 86–88)
- ❏ 1 set of Acid Rain Clue Cards (masters on pages 82–85)
- ❏ a paper clip

## For each student:
- ❏ their Environmental Detective Notebooks from previous sessions
- ❏ 1 copy of the Test Results student data sheet (masters on pages 78–79)
- ❏ 1 copy of the Newspaper Interview (master on page 81)
- ❏ *(optional)* for class work or homework: 1 copy of the Acid Rain: What Do You Think? worksheet (master on page 89)

# Getting Ready

## Before the Day of the Activity

1. If your pH paper comes on a roll, you'll need to cut a 10" piece for each pair of students. If your pH paper comes in shorter strips, have several strips per pair (enough to make about ten 1" pieces) ready to distribute.

2. Use masking tape and the permanent marker to label the nine plastic cups: "Upper Rafta River," "Lower Rafta River," "Lake Adaysicle," "Fo River," "Upper Missterssippi River," "Lower Missterssippi River," "Gray Bay," "James Pond," and "Gray Area Rain."

3. Prepare the test solutions.

   a. Add about half a cup of plain distilled water to the cups labeled "Fo River," "Upper Missterssippi River," "Lower Missterssippi River," and "James Pond." Test the water in the cups with a piece of pH paper. It should be between pH 6 and pH 8. If your distilled water is lower than pH 6, add a tiny pinch of baking soda to make it more basic. Stir thoroughly before measuring the pH.

*Note: Giving students a limited amount of pH paper prevents waste. A 10" long strip (any width is fine) is enough for each pair of students to use at all ten stations. Students can tear their strip into ten 1" pieces, one for each of the tests they'll do. Keep some extra pH paper on hand just in case.*

*Because the pH of tap water varies, we recommend that you make the solutions with distilled water.*

*As you prepare the solutions, you can always adjust their pH. A tiny pinch of baking soda will make the solutions more basic; a few drops of vinegar will make them more acidic.*

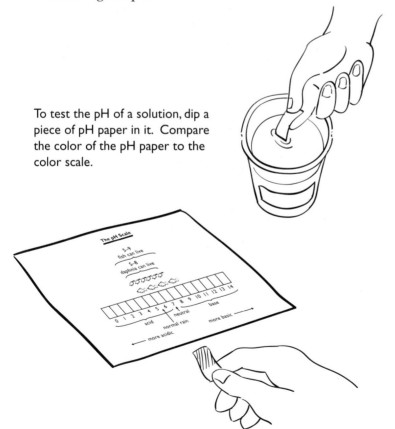

To test the pH of a solution, dip a piece of pH paper in it. Compare the color of the pH paper to the color scale.

These solutions should all be between pH 6 and 8.

*If you do add a pinch of baking soda, be sure to use clean fingers to touch the pH paper.*

b. The cups labeled "Lake Adaysicle," "Upper Rafta River," "Lower Rafta River," and "Gray Area Rain" should have a pH of about 3 or 4. Vinegar is about pH 3, so just add a little distilled water to some vinegar in each of these cups to make about half a cup of liquid in the first three cups. You'll need a full cup of Gray Area Rain.

These solutions should be pH 3 or 4.

c. The cup labeled "Gray Bay" should contain water with a pH of about 5 or 6. Add vinegar a few drops at a time to about half a cup of distilled water, and stir, until the pH is about right.

This solution should be pH 5 or 6.

4. Prepare two soil run-off models.

    a. Make a funnel and container from each of the two-liter bottles.

        1.) Cut off the top part of the bottle at a point about 1" down from where the sides begin to curve.

        2.) Turn the top piece upside down, and it becomes a funnel! Set the funnel into the bottom part of the bottle. The bottom part will serve as a large, clear container to catch the water after it drips through the funnel.

    b. Line each funnel with a coffee filter. If you're using paper towel, follow these steps:

        1.) Fold a paper towel into four quarters.

        2.) Find the corner of the folded towel which has four separate edges, then grasp the towel by the opposite corner, which will become the bottom point of the cone.

        3.) Pull one of the four separate corners away from the other three, creating a cone shape. (One side of the cone has three layers of thickness and the other has one.)

        4.) Set the filter cone into each funnel.

    c. Label one of the models "Fo and Missterssippi River Area Soil Run-off," and the other model "Rafta River Area Soil Run-off."

    d. Prepare the soils for the run-off models.

        1.) Put about half a cup of soil in a cottage cheese type container. Add 1 tablespoon baking soda and mix it. Pour this soil in the filter in the funnel of the model labeled "Fo and Missterssippi River Area Soil Run-off."

        2.) Pour about half a cup of regular soil in the filter in the funnel of the model labeled "Rafta River Area Soil Run-off."

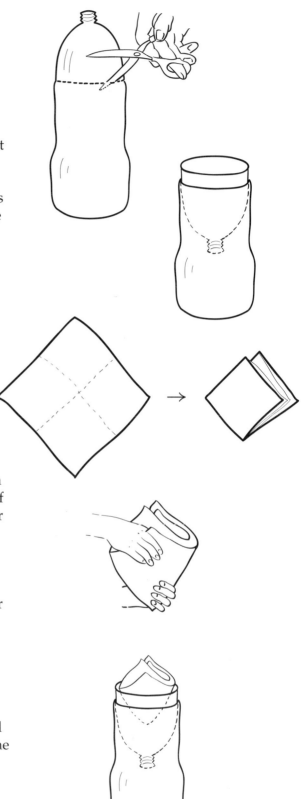

*Coloring extra pH scales can be done by a student or adult volunteer. Alternatively, you could use a color copier to make ten copies of the pH scale.*

a. To have enough colored pH scales for each of the 10 stations, first make 3 copies of the black and white pH Scale (master on page 77).

b. Using the color pH chart included with the pH paper rolls as a guide, color the pH scales with colored pencils or crayons.  Cut apart the pH scales. (You'll have 2 extra copies.)

c. If possible, laminate the pH Scales.

6. Prepare the Acid Rain Files.

*Some teachers copy the files and clue cards for each activity on a different color paper.  This way, the files for Chlorine and Acid Rain, for instance, can be easily distinguished if left out of their envelopes.  See "What You Need for the Whole Unit" on page 8.*

a. Make enough copies of the Acid Rain Files (masters on pages 86–88) so there is one set per group of four students.

b. Make enough copies of the Acid Rain Clue Cards (masters on pages 82–85) so there is one set of 12 per group of four.

c. Cut apart the clue cards, and clip them together with a paper clip.

d. Put the clue cards and file pages in an envelope for each group.

e. Label each envelope "Acid Rain Files."

7. Prepare the Acid Rain Questions.  The four questions on page 80 need to be posted for the whole class to see during the second class session.  Copy them on a piece of butcher paper or on an overhead transparency, or write them on a part of the chalkboard you can cover up until the second session.

8. Copy the Test Results sheets.  Make enough copies of the Test Results student data sheet (masters on pages 78–79) for each student to have one.  (Students will keep them in their Environmental Detective Notebooks, and use them again in later activities.)

9. Make one copy of the Newspaper Interview (master on page 81) for each student.

10. If you've decided to use them, make one copy of the Acid Rain: What Do You Think? worksheet (master on page 89) for each student.

## On the Day of the Activity

1. Keep the cup labeled, "Gray Area Rain" handy where you will introduce the first session. Copy and have on hand the suspect picture and Acid Rain statement for LaToya Faktorie (master on page 72), and Juan Tunó's and Don Juan Tunó's Acid Rain statements (masters on pages 71 and 73).

2. Set up the ten testing stations on counters or tables around the room.

   a. Place the eight labeled cups of solution you prepared earlier at eight of the stations.

   b. Make eight copies of the pH Procedure Sheet (master on page 74) and put one next to each of the eight cups.

   c. Place each of the two run-off models in separate locations (the final two stations).

   d. Make a copy of the Rafta River Soil Run-off Procedure Sheet (master on page 76) and put it next to the Rafta model. Also make a copy of the Fo and Missterssippi River Soil Run-off Procedure Sheet (master on page 75) and put it next to the other model.

   e. Place one copy of the colored pH Scale at each of the 10 stations.

   f. Have sufficient paper bags or garbage cans around the room for disposal of used pH paper.

## Session 1: Introducing Acid Rain

### Reflecting on the Evidence About Chlorine

1. As a quick review, ask the class whether they think chlorine from Ken Unballe's water slide is killing the fish and why. [Some students may be convinced that the water slide is the problem.]

2. Ask, "What do we know for sure from studying the Chlorine Files?" [There are fewer daphnia downstream from the water slide than upstream, and daphnia are sensitive to chlorine. Also, Ken Unballe wrote some pretty suspicious e-mails!] Emphasize that this is factual evidence.

3. Help students make a distinction between factual evidence and their conclusions or *inferences*. Ask, "Do we know for sure that chlorine is killing the daphnia?" "Or the fish?" [No] If we think chlorine is killing the daphnia, that is an **inference**. Inferences are good guesses, based on evidence. Tell the students it's important to make inferences, but it's also important to be ready to change your mind if new evidence comes along.

### Introducing the Acid Rain Problem in the Gray Area

1. Tell the class that Juan Tunó is still worried about chlorine, but he has concerns about acid rain, too. Read Juan's statement out loud, or have a student do so, then tape it below his picture at Headquarters.

---

**Juan Tunó**
**Acid Rain Statement**

I'm worried that **acid rain** may be killing the fish in the area. I've collected rain from near each of the three rivers. I've also collected samples to test from the Fo River, Missterssippi River, Rafta River, James Pond, Lake Adaysicle, and Gray Bay. If they're a little bit acidic, then that's normal, but if they're too acidic, it would kill the fish and other water life. I also know that rain water seeps through the soils on its way into these lakes and streams, so I also want to test water that seeps through the soils.

---

2. Hold up the cup of Gray Area Rain and say all the samples of rain Juan gathered turned out to be identical in terms of **acidity.**

3. Hold up a piece of pH paper, and explain that it tests if a substance is acidic or not. Dip it in the Gray Area Rain, and match the color of the test strip as close as possible to the color chart. (The color should indicate that the pH is about 3 or 4.) Explain that the color scale is called the pH scale, and that numbers on the color scale below seven are acid.

*If students haven't used pH paper before, emphasize that all they need to do is dip it; it isn't necessary to soak the paper for a long time.*

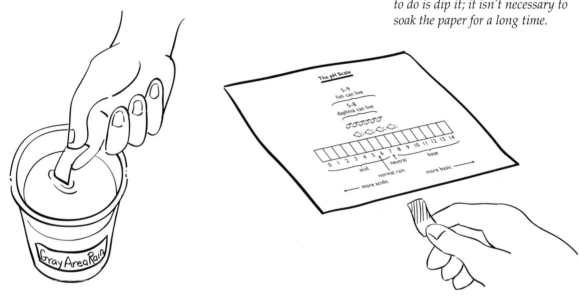

4. Tell students that the lower the number, the stronger the acid, so the Gray Area Rain is pretty acidic. Mention that seven is neutral, numbers from 0 to 6 are acidic, and numbers from 8 to 14 are called basic.

*Lemon juice and vinegar are usually about a pH of 3. Battery acid is pH 1.*

## Introducing the Tests for Acidity in Gray Area Waterways

1. Ask students if they think acid rain may be making Gray Area rivers and lakes too acidic for fish. Tell them they can find out today by going with a partner to testing stations set up around the room.

2. Point out the eight stations for tests on liquid samples from the rivers and lakes. Tell them the procedure sheet at each station will tell them how to do the test.

3. Also point out the two stations where they will test soil run-off. Explain that some acid rain falls first on the

*For this activity we chose to use vinegar (acetic acid) to represent acid rain. It's easy to obtain and is the right acidity, but is chemically different from (nitric or sulfuric) acid rain.*

*Mixing vinegar and baking soda causes a chemical reaction releasing carbon dioxide gas. If your students notice the soil "fizzing" as you pour the Gray Area Rain into the Fo and Missterssippi funnel, you may choose to discuss what's happening.*

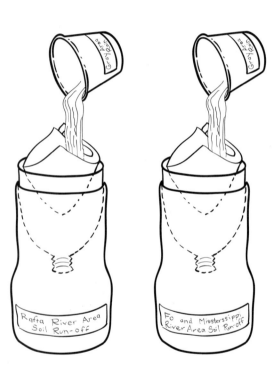

ground and then seeps through the soils into the rivers and lakes. Explain that at the stations there is soil from the Rafta and Fo/Missterssippi River areas.

4. Pour half of the cup of Gray Area Rain onto the dirt in each filter. Say that when students go to the station, some of the water that seeped through the soil will be in the bottom of the bottle for them to test. Point out the procedure sheets at the two stations.

5. Say you'll give each pair of students a strip of pH paper. They'll need to tear it into 10 little pieces, one for each of the testing stations. Ask students to put used pH paper in the paper bags or garbage cans.

6. Show the Test Results data sheet, which they will add to their detective notebooks. These sheets will be used to record their 10 test results today as well as results of future tests. Point out the section for acid rain test results.

7. Tell them they can go to the stations in any order. When they have recorded the results of each test, they will move on to any other available station. You may choose to assign each pair to their first station to make sure they're evenly distributed.

8. Tell students that when they finish testing, they can study the map and other information at Headquarters and/or their detective notebooks.

9. Distribute the pH paper and the Test Results data sheets, and have students begin.

10. When all groups have finished, give them the interview from the *Synchrony City Chronicle*. Have them read the interview with Juan Tunó now or for homework.

11. Dismantle the stations, or tidy them up if they are to be used by another class.

# Synchrony City Chronicle

Tuesday, January 23, 2001

# Gray Area Mystery—Thousands of Fish Die

## Investigators try to explain continuing fish kill

*By Florence Waters*
CHRONICLE STAFF WRITER

Concern is mounting over the major fish die-off striking the Gray Area. Investigators believe one possible explanation is acid rain. "Based on our tests, the Gray Area rain is definitely acidic," says one of the investigators, Juan Tunó. Tunó and fellow scientists are checking water samples throughout the area for acidity. What has perplexed scientists is that some rivers and lakes tested are acidic, while others are not. Juan Tunó is a student in a local school. We interviewed him in his Synchrony City home.

*Juan Tunó, student scientist*

**Thank you for your time. First of all, what exactly is an acid?**
Tunó: Acids are a group of chemicals with certain properties. They taste sour, dissolve metals, and conduct electricity well. Some examples of acids are vinegar, tomatoes, orange juice, coffee, tea, battery acids, and soda. Many acids are dangerous to taste or touch.

**How can you measure how acidic something is?**
Tunó: Scientists use something called the pH scale. The pH scale goes from 0 to 14. A chemical's pH number tells whether a chemical is acid, neutral, or base. The smaller the number, the stronger or more concentrated the acid. Vinegar is pH 3. Battery acid is pH 1.

If something is pH 7, it's neutral, and anything pH 8 and above is a base. The bigger the number, the stronger or more concentrated the base.

**Okay, so how acidic is acid rain on the pH scale?**
Tunó: Normal rain usually has a pH between 5.6 and 6. While this is a little bit acidic, it is still healthy for soils and plants. Acid rain is more acidic, and has a pH below 5.6. Some acid rain is pH 3.

**What kinds of acids are in acid rain?**

Tunó: The two main kinds of acids in rain are **nitric acid** and **sulfuric acid.**

**Wow! How do nitric and sulfuric acid get in the rain?**
Tunó: Car exhaust and factory smoke chemically react with water in clouds.

Nitric acid rain is caused by nitrogen oxides, mostly from cars and trucks. Some also comes from power plants and factories.

Sulfuric acid rain is caused by sulfur oxides. In the U.S., about 2/3 of the sulfur oxides come from coal burning and other power plants that make electricity and about 1/3 from oil refineries, factories, home heating, smelters, volcanoes, and cars.

**Well, which kind of acid rain do we have in the Gray Area?**
Tunó: We're doing some more tests on the Gray Area rain, and we hope to know soon whether cars or factories are the problem.

**What can be done?**
Tunó: We're still looking into it, but one possible suggestion would be for local factories to put "scrubbers" on their smokestacks to cut down on pollution. And we should all try to cut down on our driving.

## Session 2: Research and Discussion

### Discussing Acid Rain and the Results of pH Tests

1. Ask a few questions to review what students learned from the interview with Juan Tunó in the *Synchrony City Chronicle*:

> a. What's the pH of normal rain? [slightly acidic; about 5.6 to 6]

> b. What are the two main kinds of acid in acid rain [sulfuric and nitric acids]

> c. What causes acid rain? [Acid rain with a high level of **nitric acid** is caused mostly by cars and trucks. Some also comes from power plants and factories. Acid rain with a high level of **sulfuric acid** comes mostly from power plants, but also from oil refineries, factories, home heating, smelters, volcanoes, and cars.]

2. Ask students the pH level of the Gray Area Rain. [about pH 3] Ask, "Is that normal or acid rain?" [acid rain]

*If some students have differing test results, spend a few minutes listing the possible reasons for this. [Different interpretations of the pH scale, possible contamination of some samples, etc.] Variations in test results often happen to scientists too, and sometimes lead to re-testing.*

3. Tell students that most fish can't survive in water below pH 5. Have them look in their notebooks for their pH test results. Ask, "Do any of the rivers or lakes you tested have a pH level that is too acidic for fish to survive?" [The Upper and Lower Rafta River and Lake Adaysicle are too acidic for fish (with a pH of about 3). The soil run-off from the Rafta River area is also too acidic for fish.]

4. Ask, "Were there any water samples that were **not** too acidic for fish?" [The Fo River, Upper Missterssippi, Lower Missterssippi, Gray Bay, and James Pond have a pH of about 5 or higher, which is not too acidic for fish. The soil run-off from the Fo and Missterssippi areas is also not too acidic for fish.]

5. Take a few minutes for students to share their ideas and questions about their test results.

### Small Group Research and Discussion

1. Post the four Acid Rain Questions you prepared earlier. Tell the class that doing some research in the Acid Rain Files will help them answer these four questions.

2. Say that each group of four will get an Acid Rain File for their research. Tell the class the files contain quite a few documents and also clue cards with information about acid rain.

3. Tell students that groups can divide up some of the reading as they did in the Chlorine Files. Since there are four acid rain questions, suggest that, this time, each student in a group be responsible for seeking the answer to one question.

4. Suggest that they all *skim* the documents to "inventory" what's in them, and then put them in different piles according to which of them may help answer each of the questions. Caution them that some of the documents may not fit into easy categories.

### "Skimming 101"

*Skimming is an important research skill, especially when there are many documents, as in the Acid Rain Files. If you think your students need tips on how to skim, you might want to practice together before they start their research. Make a transparency or a chalkboard version of a clue card, such as this one:*

**Are taller smokestacks the solution to acid rain?**

In the 1960s, some factories were getting complaints from people near the factories, so they came up with a "solution." They made their smokestacks taller, and the pollutants blew far away in the wind, even to other countries. But, before too long, these factories received complaints for causing acid rain problems in other places.

*Here are some tips on how to skim this clue card to see if it's going to be useful:*

- *Keep your question in mind as you read. [Say, for example, your research question is #4 "What are some possible solutions to the acid rain problem?"]*

- *Read the title. [It looks promising.]*

- *Look for key words in the text like "solution" or "solve."*

- *Don't take time to figure out every word. [If you don't know the word pollutant, it's okay; you get the idea that the smoke is blowing farther away.]*

- *Get the general idea. [This card is about how taller smokestacks could reduce local acid rain.]*

- *Don't try to learn details now. [Even if it's interesting, don't slow down now to read carefully and start thinking about it.]*

*It may be challenging for students to determine which documents are relevant to the questions under investigation. For example, the clue cards about the geology around waterways may not seem relevant at first, but they are key to understanding why some waterways are more acidic than others. Encourage students who are having trouble answering a question to review the documents in their "extra" pile.*

5. Say that some documents may turn out to be interesting but not helpful in answering any of the questions! That's just the way research is. Suggest they make an "extra" pile for documents that don't seem to be related directly to any of the questions.

6. Each student should take notes in their notebook as they do research on their question. If a student finishes quickly, they should help a teammate.

7. When all four students are ready, each will share and discuss their findings with the group.

8. Have one student from each group get an Acid Rain File, and have groups begin their research.

## Large Group Discussion

1. When groups have finished their research, use the four Acid Rain Questions to facilitate a class discussion. (It is likely that the class will agree with the responses in brackets below. **However, it is very important to accept all answers, and to ask students to share their reasoning.**)

2. Ask the first question, "What do you think causes an acid problem in some lakes and streams, but not others?" [The Upper and Lower Rafta River and Lake Adaysicle are surrounded by granite. The other bodies of water are surrounded by limestone, which naturally buffers acid.]

3. Next ask, "Do you think acid rain is killing the fish?" [probably] "Why or why not?" [Acidity below pH 5 kills some fish, and is harmful to their eggs.]

4. Then ask, "Who or what do you think is causing acid rain in the Gray Area?" "Why do you think so?" [The area's acid rain is mostly nitric acid, which is generally caused mostly by cars and trucks.]

5. Finally ask, "What are some possible solutions to the acid rain problem?" [Putting scrubbers on factory smokestacks, adding buffers like limestone to lakes, making cleaner cars, driving less...]

6. Also ask students if they think acid rain or chlorine is killing the water fleas near the water slide? [It's not acid rain, because the Fo River isn't acidic. It may be chlorine.]

7. Ask if students found out anything else interesting in the files.

8. Show the picture of LaToya Faktorie, have a student volunteer read her statement, then tape it on the Suspect Chart.

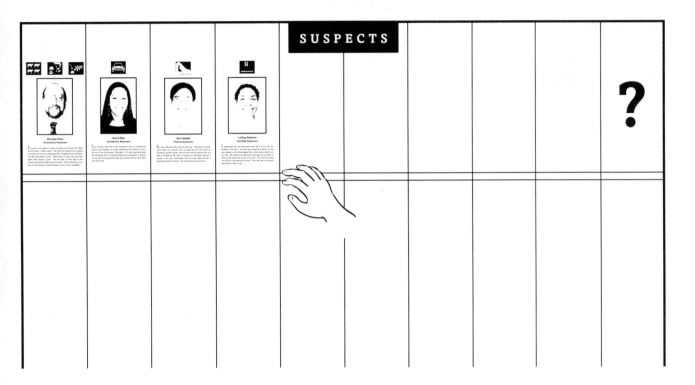

## LaToya Faktorie
### Acid Rain Statement

I understand that you discovered that there is an acid rain problem in the area. Your tests also proved that there's no acid rain problem in the Missterssippi River, which proves that it's not our fault. We received an award three years ago for our efforts to clean up the waste that we put in the river. The cars are causing the acid rain, and that's not our fault. That's the fault of everyone who drives or rides in cars.

9. Ask students if they have any comments or reactions to her statement, or if they disagree with anything she said.

10. Ask a student volunteer to read the following statement by Don Juan Tunó, owner of the oil refinery, and then tape it below his picture and previous statement on the Suspect Chart.

### Don Juan Tunó
#### Acid Rain Statement

We release sulfur oxides into the air, but we do it to make products that you buy and use. If you choose to close down oil refineries, then you might as well give up cars, jet skis, natural gas, fertilizers, wax, chewing gum, perfumes, cloth, plastics, asphalt, and a whole lot of other stuff. As long as you keep using these products, you'll need us around.

11. Ask students if they have any comments or reactions to his statement, or if they disagree with anything he said.

## Reflect, Predict, and Vote

1. Tell students they will each receive two post-its, and they'll write their names in large letters on each. They will use these post-its to vote on which of the suspects they think might be "guilty" of causing the fish kill. They'll put their post-it on the Suspect Chart under the names of the two people they suspect most.

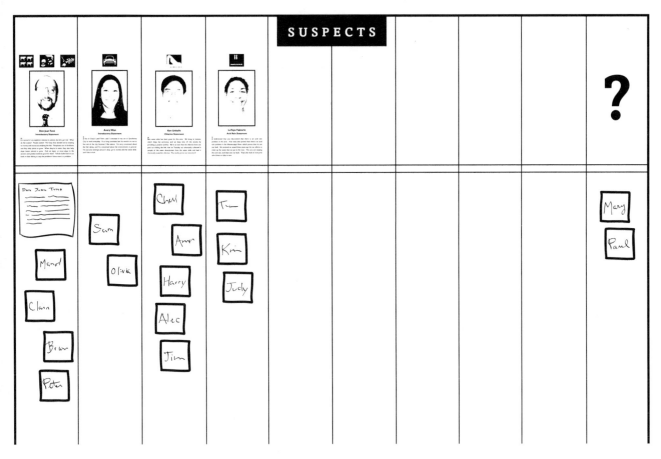

2. Point out that there is also a column on the chart with a question mark. They can put post-its there if they are undecided.

3. Let students know they can change their votes at any time during the unit, and not to worry about making sure they've got the "right" answer. Explain that the purpose of the post-it vote is mostly to see how the ideas of the class change over time—so all their participation is needed.

4. Tell your students they will have a few minutes to:

- Write notes in their detective notebooks.

- Write "acid rain" on the map in their detective notebooks where they think acid rain is coming from.

- Put post-its under suspects on the chart.

5. If you've decided to use it, assign as homework (or have students work in class on) the Acid Rain: What Do You Think? worksheet. Have students put this sheet in their Environmental Detective Notebooks.

## Going Further

**1. Acid Rain Research in your Area.**

a. Have students collect newspaper articles or other reports on current events that relate to acid rain.

b. Encourage students to collect rain from their homes and check the pH. They could also enlist friends in different geographical locations to get more data. Where in your area is acid rain most prevalent? This could also make an excellent Science Fair project.

c. Suggest that students collect different waters—from the drinking fountain, ponds, streams, rain, water with detergent, and/or water with decomposing material in it—and test the pH. Water fit to drink (pH-wise) can range from 6.5–8.5 pH.

**2. More Acid Rain Activities.** The GEMS *Acid Rain* guide could be used for pre-teaching or as a follow-up to these activities. The unit explores many aspects of acid rain and its effects. Students gain scientific inquiry skills as they make "fake lakes," determine how the pH of the lakes changes after an acid rainstorm, present a play about the effects of acid rain on aquatic life, and hold a town meeting to discuss solutions to the problem. Students also play a "Startling Statements" game and conduct a plant growth experiment. Extensive background information for the teacher is also provided. If you don't have time for the whole unit, you may choose to just do the acid rain play as a supplement to this activity.

**3. More pH Activities.** The GEMS guide *Of Cabbages and Chemistry* could also be used for pre-teaching or as a follow-up. The unit offers students a chance to explore acids and bases using the special indicator properties of cabbage juice. Students discover that chemicals can be grouped by behaviors, and relate acids and bases to their own daily experience. The color-change game Presto Change-O helps students discover the acid-neutral-base continuum. The guide includes an "Acid and Aliens from Outer Space" assessment. *Of Cabbages and Chemistry* is also an excellent lead-in to the GEMS guide *Acid Rain*.

# Juan Tunó
## Acid Rain Statement

I'm worried that **acid rain** may be killing the fish in the area. I've collected rain from near each of the three rivers. I've also collected samples to test from the Fo River, Missterssippi River, Rafta River, James Pond, Lake Adaysicle, and Gray Bay. If they're a little bit acidic, then that's normal, but if they're too acidic, it would kill the fish and other water life. I also know that rain water seeps through the soils on its way into these lakes and streams, so I also want to test water that seeps through the soils.

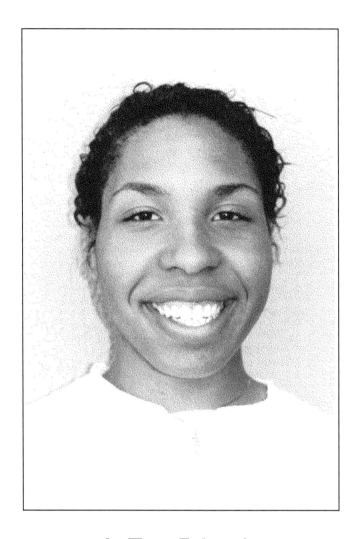

## LaToya Faktorie
### Acid Rain Statement

I understand that you discovered that there is an acid rain problem in the area. Your tests also proved that there's no acid rain problem in the Missterssippi River, which proves that it's not our fault. We received an award three years ago for our efforts to clean up the waste that we put in the river. The cars are causing the acid rain, and that's not our fault. That's the fault of everyone who drives or rides in cars.

# Don Juan Tunó
## Acid Rain Statement

We release sulfur oxides into the air, but we do it to make products that you buy and use. If you choose to close down oil refineries, then you might as well give up cars, jet skis, natural gas, fertilizers, wax, chewing gum, perfumes, cloth, plastics, asphalt, and a whole lot of other stuff. As long as you keep using these products, you'll need us around.

**1.** Dip a piece of pH paper in the test solution.

**2.** Match the color of the wet paper with the colors on the pH scale.

**3.** On your Test Results data sheet, mark down the pH number and whether it is OK for fish, too acidic for fish, or too basic for fish.

**4.** Throw away your used pH paper in the garbage can.

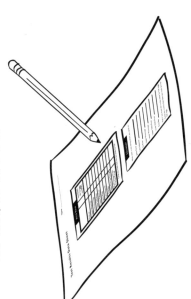

**1.** Take out the top half of the bottle.

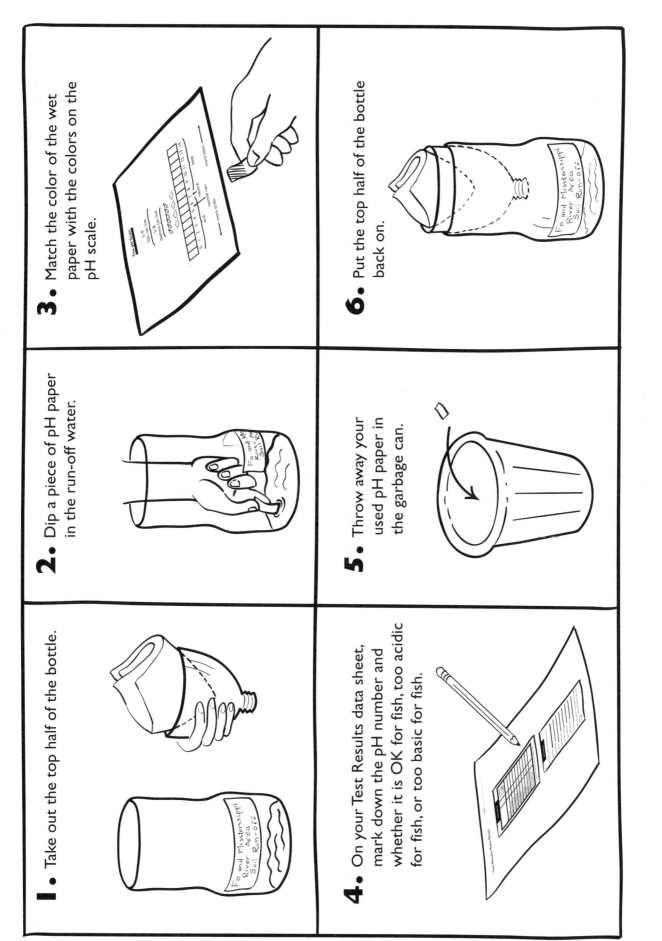

**2.** Dip a piece of pH paper in the run-off water.

**3.** Match the color of the wet paper with the colors on the pH scale.

**4.** On your Test Results data sheet, mark down the pH number and whether it is OK for fish, too acidic for fish, or too basic for fish.

**5.** Throw away your used pH paper in the garbage can.

**6.** Put the top half of the bottle back on.

**1.** Take out the top half of the bottle.

**2.** Dip a piece of pH paper in the run-off water.

**3.** Match the color of the wet paper with the colors on the pH scale.

**4.** On your Test Results data sheet, mark down the pH number and whether it is OK for fish, too acidic for fish, or too basic for fish.

**5.** Throw away your used pH paper in the garbage can.

**6.** Put the top half of the bottle back on.

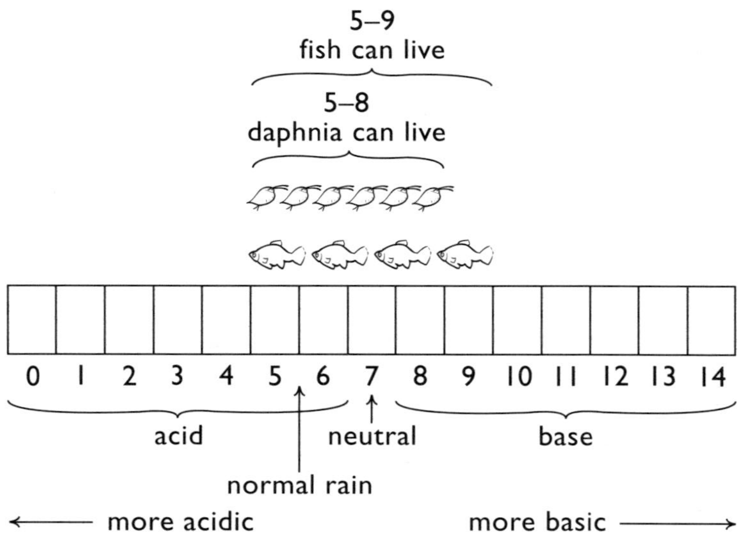

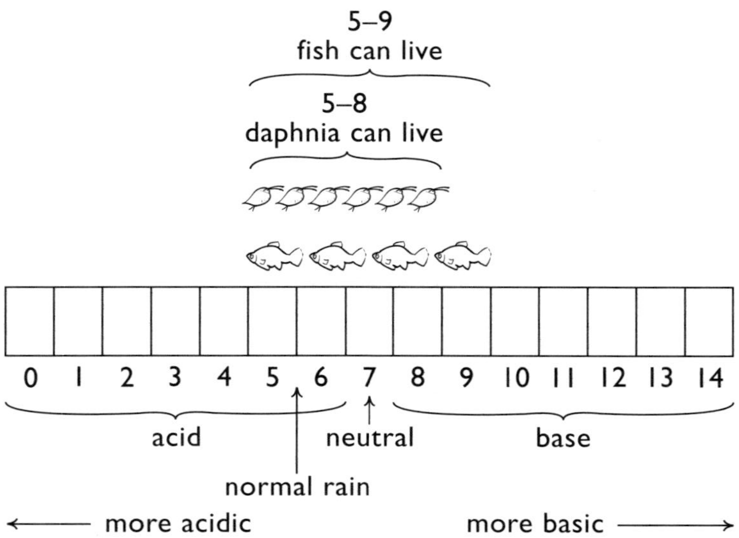

The pH Scale

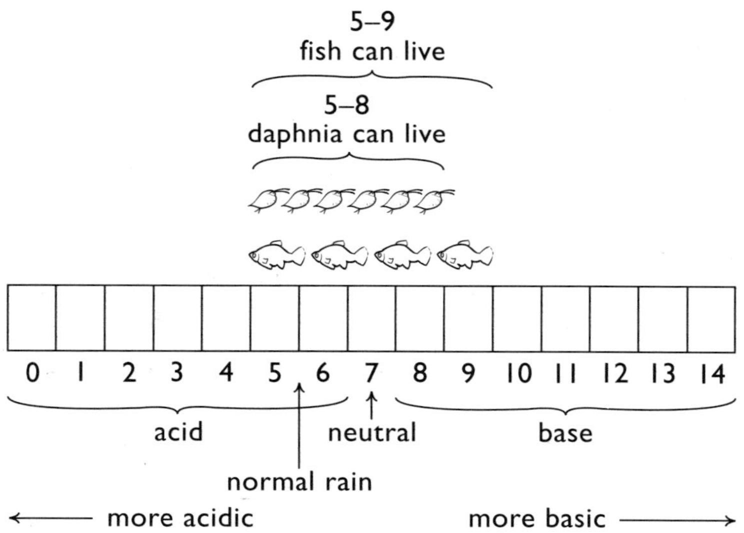

5–9
fish can live

5–8
daphnia can live

0 1 2 3 4 5 6 7 8 9 10 11 12 13 14

acid    neutral    base

normal rain

←— more acidic        more basic —→

---

The pH Scale

5–9
fish can live

5–8
daphnia can live

0 1 2 3 4 5 6 7 8 9 10 11 12 13 14

acid    neutral    base

normal rain

←— more acidic        more basic —→

---

The pH Scale

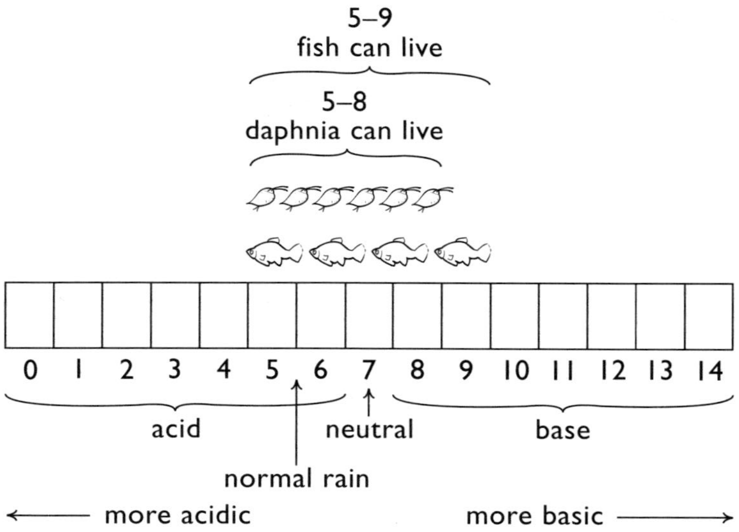

5–9
fish can live

5–8
daphnia can live

0 1 2 3 4 5 6 7 8 9 10 11 12 13 14

acid    neutral    base

normal rain

←— more acidic        more basic —→

---

The pH Scale

5–9
fish can live

5–8
daphnia can live

0 1 2 3 4 5 6 7 8 9 10 11 12 13 14

acid    neutral    base

normal rain

←— more acidic        more basic —→

# Test Results Data Sheet

Name _____

## ACID RAIN

|  | pH # | too ACIDIC, OK for fish, too BASIC |
|---|---|---|
| Fo River |  |  |
| Upper Missterssippi River |  |  |
| Lower Missterssippi River |  |  |
| Upper Rafta River |  |  |
| Lower Rafta River |  |  |
| James Pond |  |  |
| Lake Adaysicle |  |  |
| Gray Bay |  |  |
| Rafta River Area Soil Run-off |  |  |
| Fo/Missterssippi River Area Soil Run-off |  |  |

Circle the rivers that are too acidic for fish.

## SEDIMENT

What color is the string where you pinch it?

James Pond _____

Lake Adaysicle _____

Fo River _____

Upper Missterssippi River _____

Lower Missterssippi River _____

Upper Rafta River _____

Lower Rafta River _____

Gray Bay _____

Black = too much dirt in the water for fish eggs.

Circle the river(s) that have too much dirt for fish eggs.

# Test Results Data Sheet – continued

**WATER LIFE**

Write the names of the water life found in James Pond:

_____

_____

_____

_____

_____

_____

Is James Pond

**HEALTHY**

or

**TOO LOW IN OXYGEN?**

**PHOSPHATES**

Are phosphate levels HIGH or OK?

Cattle Ranch Drainage into Pond _____

Golf Course Drainage into Pond _____

Small Town Drainage into Pond _____

Circle the places

with too many

phosphates.

**BIRDWATCHING**    K = Kingfisher • H = Hawk • D = Duck

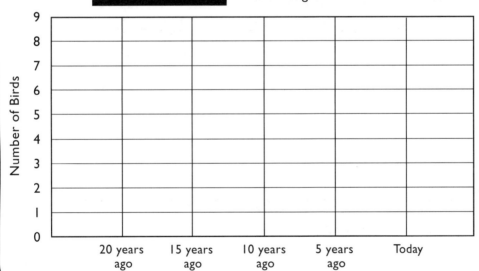

**OIL**    Oil in the Ocean

Circle the place where the oil is coming from:

tanker        car        refinery

# Acid Rain Questions

1. What do you think causes an acid problem in some lakes and streams, but not others?

2. Do you think acid rain is killing the fish? Why or why not?

3. Who or what do you think is causing acid rain in the Gray Area? Why do you think so?

4. What are some possible solutions to the acid rain problem?

Tuesday, January 23, 2001

# Gray Area Mystery—Thousands of Fish Die

## Investigators try to explain continuing fish kill

*By Florence Waters*
CHRONICLE STAFF WRITER

Concern is mounting over the major fish die-off striking the Gray Area. Investigators believe one possible explanation is acid rain. "Based on our tests, the Gray Area rain is definitely acidic," says one of the investigators, Juan Tunó. Tunó and fellow scientists are checking water samples throughout the area for acidity. What has perplexed scientists is that some rivers and lakes tested are acidic, while others are not. Juan Tunó is a student in a local school. We interviewed him in his Synchrony City home.

*Juan Tunó, student scientist*

**Thank you for your time. First of all, what exactly is an acid?**
Tunó: Acids are a group of chemicals with certain properties. They taste sour, dissolve metals, and conduct electricity well. Some examples of acids are vinegar, tomatoes, orange juice, coffee, tea, battery acids, and soda. Many acids are dangerous to taste or touch.

**How can you measure how acidic something is?**
Tunó: Scientists use something called the pH scale. The pH scale goes from 0 to 14. A chemical's pH number tells whether a chemical is acid, neutral, or base. The smaller the number, the stronger or more concentrated the acid. Vinegar is pH 3. Battery acid is pH 1.

If something is pH 7, it's neutral, and anything pH 8 and above is a base. The bigger the number, the stronger or more concentrated the base.

**Okay, so how acidic is acid rain on the pH scale?**
Tunó: Normal rain usually has a pH between 5.6 and 6. While this is a little bit acidic, it is still healthy for soils and plants. Acid rain is more acidic, and has a pH below 5.6. Some acid rain is pH 3.

**What kinds of acids are in acid rain?**

Tunó: The two main kinds of acids in rain are **nitric acid** and **sulfuric acid.**

**Wow! How do nitric and sulfuric acid get in the rain?**
Tunó: Car exhaust and factory smoke chemically react with water in clouds.

Nitric acid rain is caused by nitrogen oxides, mostly from cars and trucks. Some also comes from power plants and factories.

Sulfuric acid rain is caused by sulfur oxides. In the U.S., about 2/3 of the sulfur oxides come from coal burning and other power plants that make electricity and about 1/3 from oil refineries, factories, home heating, smelters, volcanoes, and cars.

**Well, which kind of acid rain do we have in the Gray Area?**
Tunó: We're doing some more tests on the Gray Area rain, and we hope to know soon whether cars or factories are the problem.

**What can be done?**
Tunó: We're still looking into it, but one possible suggestion would be for local factories to put "scrubbers" on their smokestacks to cut down on pollution. And we should all try to cut down on our driving.

## What pH can water life live in?

| | most acidic | neutral | most basic |
|---|---|---|---|
| | 0 1 2 3 4 5 6 7 8 9 10 11 12 13 14 | | |

~ pH 2–13
Bacteria

~ pH 6.5–12.5
Plants (algae, rooted, etc.)

~ pH 5–9
Carp, suckers, catfish, some insects

~ pH 6–9.5
Bass, bluegill, crappie

~ pH 7.5–10
Snails, clams, mussels

~ pH 7–9
Largest variety of animals (trout, mayfly nymphs, stonefly nymphs, caddisfly larvae)

ACID RAIN CLUE CARD • ACID RAIN CLUE CARD • ACID RAIN CLUE CARD • ACID RAIN CLUE CARD • ACID RAIN CLUE CA

## What is the effect of acid rain on fish?

If water becomes too acidic, some fish survive, but many others die. Even if the adult fish live, they may not be able to have babies.

acid rain

ACID RAIN CLUE CARD • ACID RAIN CLUE CARD • ACID RAIN CLUE CARD • ACID RAIN CLUE CARD • ACID RAIN CLUE CA

## What is the effect of acid rain on water plants?

If the water is too acidic, water plants (algae) are not able to take in nutrients as well as they should. Their growth is stunted, and eventually they die.

acid rain

ACID RAIN CLUE CARD • ACID RAIN CLUE CARD • ACID RAIN CLUE CARD • ACID RAIN CLUE CARD • ACID RAIN CLUE CA

## [A]ren't healthy lakes acidic?

[M]ost lakes are a little bit acidic, but the [la]kes with the most life in them are a little [bi]t basic. Water plants can make more [o]xygen in lakes that are slightly basic. [W]ater life needs oxygen in the water to [b]reathe.

a little bit acidic

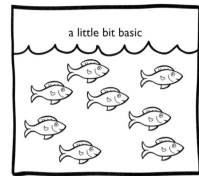

a little bit basic

ACID RAIN CLUE CARD • ACID RAIN CLUE CARD • ACID RAIN CLUE CARD • ACID RAIN CLUE CARD • ACID RAIN CLUE CARD

## [H]ow does acid rain affect soil?

[W]hen acid rainwater goes through soil, it can [c]hange the soil's pH and wash away nutrients. [T]he tiny animals that need the nutrients die, [t]hen larger animals that feed on them die.

acid rain

ACID RAIN CLUE CARD • ACID RAIN CLUE CARD • ACID RAIN CLUE CARD • ACID RAIN CLUE CARD • ACID RAIN CLUE CARD

## [H]ow can soil change the pH of acid [r]ain?

[W]hen acid rain falls on some kinds of soil, [i]t can be *neutralized* (made not acid). This [i]s also called *buffering* the acid. Limestone [a]nd clay are two soils that can neutralize or [b]uffer acids. Granite soil does not neutral[i]ze acids.

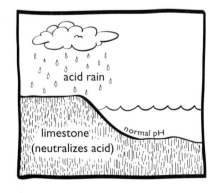

acid rain

limestone (neutralizes acid)    normal pH

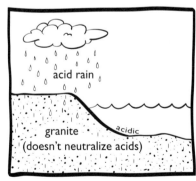

acid rain

granite (doesn't neutralize acids)    acidic

ACID RAIN CLUE CARD • ACID RAIN CLUE CARD • ACID RAIN CLUE CARD • ACID RAIN CLUE CARD • ACID RAIN CLUE CARD

## Geology of the Rafta River area

The soils and rocks around the Rafta River are mostly granite. Granite is an igneous rock that does not neutralize acids.

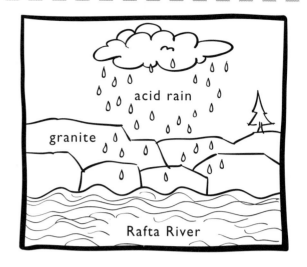

ACID RAIN CLUE CARD · ACID RAIN CLUE CARD · ACID RAIN CLUE CARD · ACID RAIN CLUE CARD · ACID RAIN CLUE CA

## Geology of the Fo and Missterssippi River area

The soils and rocks around the Fo and Missterssippi Rivers are mostly limestone. Limestone is a sedimentary rock that neutralizes acids.

ACID RAIN CLUE CARD · ACID RAIN CLUE CARD · ACID RAIN CLUE CARD · ACID RAIN CLUE CARD · ACID RAIN CLUE CA

## What are scrubbers and how can they help prevent acid rain?

Scrubbers clean factory smokestacks, so they will release fewer chemicals that cause acid rain. They are expensive to buy, but easy to maintain. They stop acid rain from ever forming.

ACID RAIN CLUE CARD · ACID RAIN CLUE CARD · ACID RAIN CLUE CARD · ACID RAIN CLUE CARD · ACID RAIN CLUE CAR

## taller smokestacks the solution to acid ...?

In the 1960s, some factories were getting complaints from people near the factories, so they came up with a "solution." They made their smokestacks taller, and the pollutants blew far away in the wind, even to other countries. But, before too long, these factories received complaints for causing acid rain problems in other places.

## How are buffers used to solve acid rain problems?

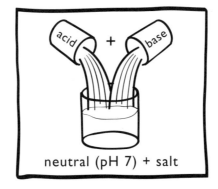

Buffers, such as limestone, can be added to acid lakes to make them less acidic. This is called "liming" the lake. It is only a temporary and expensive solution because the lake must be "limed" every year as long as there is acid rain falling.

## What is neutralization?

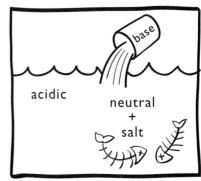

When an acid is added to a base it can make it neutral (pH 7). Bases can also neutralize acids. One drawback when bases are used to neutralize acids in acid lakes is that certain salts are also produced which can be bad for freshwater lakes and streams.

# Gray Area's Gray Air

### Gray Area listed as severe by Environmental Protection Agency

*By Kimi Bevilacqua*
PERMANENT PRESS

The Gray Area is one of thirty-three regions in the country that have not cleaned up their air enough to meet federal smog standards, the Environmental Protection Agency (EPA) said today. Last year the Gray Area was listed as "serious," but this year it has been changed to "severe," because it has gotten worse. Scientists blame local polluters, as well as pollution carried by the winds from other areas.

Only one city in the country is now ranked as "extreme." For most people in the United States, the air is cleaner than it was five years ago because of the EPA's efforts.

# Local Factory Releasing Smoke at Night

WEDNESDAY, JANUARY 24, 2001

### "It's not a plot!" says LaToya Faktorie

*By Lincoln Roston*
KEEPA JOURNAL

The local toy factory has been releasing many of their pollutants at night. During the day, the factory mainly releases white smoke, which can be mistaken for steam, but they save any dark pollutants to release at night, when it's harder to see them. "We only release the darker smoke at night, because that's when our factory has been set up to do the work that makes it," said owner LaToya Faktorie. "It's not a plot!"

# Car Makers Agree to Make Cleaner Cars

*By Vivian Beals*
PICKIER NEWS

The six biggest car manufacturers in the world have agreed to make cars that pollute less. The added cost per car will probably be about $95. Two states have demanded that the manufacturers also be required to make 2% of their cars electric or zero-emission vehicles.

# Synchrony City Chronicle
**Tuesday, January 23, 2001**

## Letters to the Editor

Editor:

I love my car, and you're not going to get me to take the bus.

In your newspaper last week, you said that air pollution from our cars and factories is making acid rain and killing our fish. But that can't be right!

Here's why: If acid rain is falling on the Rafta River, then it must be falling on the Fo and Missterssippi Rivers too. Acid rain makes lakes clear, because it kills living things, including algae. The Rafta looks clear, but the Missterssippi and Fo don't. Explain that to me!

Signed, Concerned Citizen

Dear Editor,

Thank you for your suggestion in last week's newspaper to put scrubbers on our smokestacks to cut down on our pollution. We want to pollute the air as little as possible, but we are a business, and we have to make a profit. The scrubbers you suggest would cut pollution, but they would cost too much money for our company. We have won an award for cleaning up our pollution of the Missterssippi River, so we really do care.

Thanks again,
LaToya Faktorie

# Gray Area Acid Rain Analysis

88% acid rain with a high level of nitric acid

12% acid rain with a high level of sulfuric acid

# Environmental Action Award

This award is to certify that the **Gray Area Toy Factory** has reduced their water pollutants by 11%. The wildlife and people of our country for decades to come will benefit from your efforts.

**Clean Water Association**

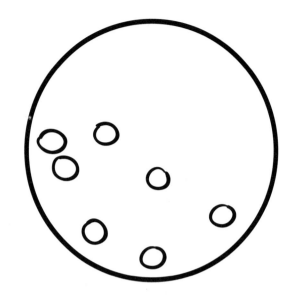

pH 5.5 water (less acid)
Brown trout eggs develop normally

pH 4.6 water (more acid)
Brown trout eggs don't develop

# Polluters That Cause Acid Rain

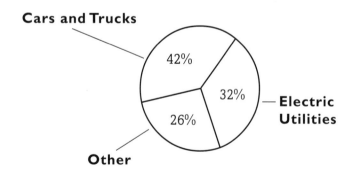

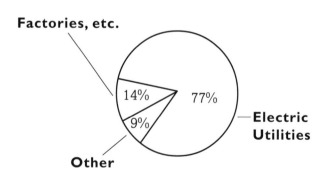

## Nitric Acid Rain

Where $NO_x$ pollution comes from
25.2 million tons per year

## Sulfuric Acid Rain

Where $SO_x$ pollution comes from
19.5 million tons per year

# Acid Rain: What Do You Think?

1. Which lakes and rivers are too acidic for fish? _____

_____

2. Which lakes and rivers are **not** too acidic for fish? _____

_____

3. Acid rain is falling on the whole Gray Area. Why aren't all the rivers and lakes acidic? _____

_____

_____

_____

4. Do you think acid rain is killing the water fleas that Juan tested? Explain. _____

_____

_____

5. Do you think acid rain is killing the fish?  Explain. _____

_____

_____

6. Who or what do you think is causing acid rain in the Gray Area?

_____

_____

_____

# Activity 4: Sedimental Journey

## Overview

In this activity, students are introduced to erosion and sedimentation in waterways through the use of a model projected on the overhead. Statements by Juan Tunó and from a new suspect—a logger—start your students wondering: Has there been a loss of vegetation in the Gray Area? Could this be causing erosion and excess sedimentation? Could that be what is killing the fish? And, if it is, *who's* to blame?

Then pairs of students circulate to eight learning stations to test the amount of sediment in samples of water from eight different Gray Area locations. Groups do research using the "Sediment Files" and hold a discussion. Students are given time to reflect, write the results of their tests and their research, and adjust their predictions on the Suspect Chart.

*During their research, students discover that sedimentation can harm aquatic plants, insects, and fish in many ways, including by darkening the water, causing the water to absorb more sunlight, thus raising the water temperature.*

*It's well worth taking the time to practice the erosion demonstration before class. Making sure you have the right pan, soil, and dribbling technique will avoid mess and frustration. When it works, this demonstration is very effective. One teacher told us: "Excellent! Students said, 'Oooh, ah!' Great visual model to clearly show erosion. Helped my kids to understand the concept." Another teacher said: "This was wonderful!! The children loved it and it was very concrete. They were quite enthused to get started with the sediment tests."*

## What You Need

### For the erosion demonstration:
❐ an overhead projector
❐ 1 *transparent* glass or plastic tray or pan at least 9" x 12" and at least 2" deep. (A sturdy container is preferable. If you use a lightweight container, such as a flat deli-style salad container, be sure to test it before class to make certain it doesn't leak or tip over easily.)
❐ 2 plastic flat-bottomed cups (about 8 or 9 oz. size)
❐ a small bowl or cup for extra water
❐ 1 pair of scissors
❐ 1 tablespoon measuring spoon or a graduated cylinder (15 ml or more)
❐ paper towels
❐ 4 or more tablespoons of soil (You need soil that will run off when "rained" on. Use any plain soil that isn't too rocky, clay-like, or too full of organic material like roots and leaves. Avoid potting soil.)
❐ a handful of one or more of the following: pencil sharpener shavings, grass clippings, moss, dried weeds, or cut up pieces of paper towel

## For the eight learning stations:

- ❏ 8 large washers
- ❏ 1 roll of masking tape, or light-colored labeling tape (preferably white)
- ❏ 8 one-foot pieces of white string
- ❏ 1 black permanent marker
- ❏ 8 milk cartons (half-gallon or quart size)
- ❏ 8 unsharpened pencils
- ❏ 1 half-pint of chocolate milk
- ❏ spoon or ruler
- ❏ 8 copies of the Sediment Test Procedure Sheet (master on page 106)
- ❏ paper towels

## For the class:

- ❏ Headquarters set-up from previous sessions
- ❏ 1 copy of Anton Alogue's picture and Sediment statement (master on page 104)
- ❏ 1 copy of Elmo Skeeto's picture and Sediment statement (master on page 105)
- ❏ 1 copy of Juan Tunó's Sediment statement (master on page 103)

## For each group of four students:

- ❏ a 9" x 12" envelope or file folder
- ❏ 1 copy of the Sediment Files (masters on pages 107–110)

## For each student:

- ❏ their Environmental Detective Notebooks from previous sessions
- ❏ 1 copy of the Sediments: What Do You Think? student sheet (master on page 111)

# Getting Ready

## Before the Day of the Activity

1. Prepare eight milk cartons for the sediment test learning stations.

    a. Use masking tape and a permanent marker to label eight milk cartons: "James Pond," "Lake Adaysicle," "Fo River," "Upper Misstersssippi River," "Lower Misstersssippi River," "Upper Rafta River," "Lower Rafta River," and "Gray Bay."

    b. Unfold the top of each milk carton so the top is completely open.

*Note: The disks you make for this activity are much smaller than the real 10" Secchi (SEH-kee) disks used to test for turbidity (how much dirt is suspended in water). The disk is lowered into water until it can no longer be seen, and then the length of rope used to lower it is measured. See page 217 for more background information on Secchi disks. For this classroom activity, students lower a simulated Secchi disk made from a washer and string into a milk carton of water made "turbid" with chocolate milk. The materials can be used by multiple classes without additional preparation.*

2. Make eight Secchi disks. To make each disk:

    a. Tie a piece of string around a washer.

    b. Use masking tape or light-colored labeling tape (ideally white) to wrap around the washer to make it look like a solid disk. While taping, hold the string so it will come out from the center of the disk. This will allow the disk to hang horizontally.

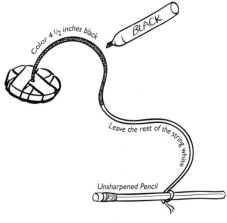

Color 4 1/2 inches black

BLACK

Leave the rest of the string white

Unsharpened Pencil

c. Tie the other end of the string to an unsharpened pencil, adjusting the length so that the washer can hang to the bottom of an opened milk carton (about 8").

d. Color about 4 ½" of string directly above the washer with the **black** permanent marker.

e. Leave the rest of the string white.

3. Make one copy each of the picture and Sediment statements for Anton Alogue (master on page 104) and Elmo Skeeto (master on page 105), and Juan Tunó's Sediment statement (master on page 103).

4. Make eight copies of the Sediment Test Procedure Sheet (master on page 106).

5. Prepare the Sediment Files (masters on pages 107–110). Make enough copies of the articles and documents so there is one set per group of four. Put each set of articles and documents in an envelope. Label each envelope "Sediment Files."

6. Make one copy of the Sediments: What Do You Think? sheet (master on page 111) for each student. In this activity, these will be used instead of discussion cards during the small group discussion.

**On the Day of the Activity**

**Preparing Materials for the Erosion Demonstration**

1. Fill the transparent pan with about 1" of water, and set it on the overhead projector.

2. Use the scissors to cut a triangle in the rim of each of the two small plastic cups. The cups represent "hills" in a model lake, and the triangular hole will allow them to sit upside down in the water.

3. Turn the cups upside down on a paper towel. Scoop up a tablespoon of soil and place it in a mound on one cup.

4. Scoop up another tablespoon of soil. Put it on the paper towel and mix it with a few pinches of grass clippings (or

*If your soil is very powdery, you may want to mist it lightly so that it will clump enough to stay put when you pile it on top of the cup.*

whatever materials you decided to use.) Repack the tablespoon, and place the mixture on top of the other cup.

5. Have handy near the overhead projector: both prepared cups, a tablespoon (or a graduated cylinder), a cup or bowl of extra water, the suspect pictures and statements for Anton Alogue and Elmo Skeeto, and Juan Tunó's statement.

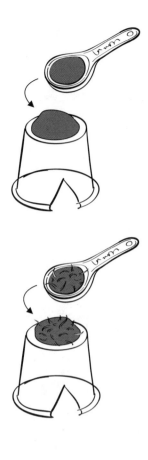

6. Practice the erosion demonstration on page 97 ahead of time.

## Setting Up the Sediment Test Learning Stations

1. Fill each milk carton about 8" deep with water.

2. Vary the murkiness of the water in different cartons by stirring small amounts of chocolate milk into the water as follows.

    a. Don't add any chocolate milk to Upper Rafta, Lower Rafta, or Lake Adaysicle. *(no sediment)*

b. Add just a few drops of chocolate milk to the Fo River, Gray Bay, and James Pond, so the disk can still be seen when it is 5" or deeper (so if you pinch the string at the surface when it disappears, you're pinching the **white** part of the string). *(medium sediment)*

*Optional:* You may want to create some variation in muddiness among these three cartons, staying within the "white" range.

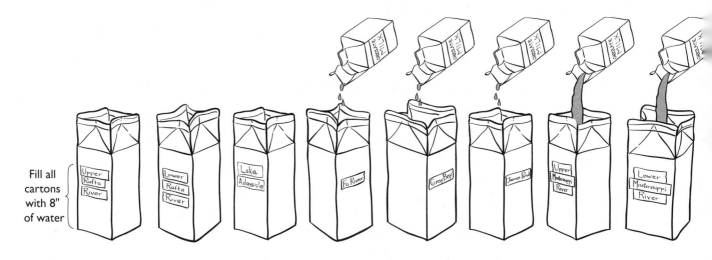

Fill all cartons with 8" of water

c. Add small amounts of chocolate milk to the Upper and Lower Missterssippi samples so, when the disk is lowered, it disappears when it is less than 4" deep (so if you pinch the string at the surface when it disappears, you're pinching the **black** part of the string). *(high sediment)*

*Optional:* You may want to make the Upper Missterssippi sample slightly more muddy than the Lower Missterssippi.

3. Choose eight locations for learning stations on counters and tables around the room as widely spaced as possible, so circulating pairs of students can gather near them.

4. Put one milk carton at each station. Stir the chocolate milk if it has begun to settle. Place a Secchi disk on a paper towel next to the milk carton. Put one copy of the Sediment Test Procedure sheet at each station.

## Introducing the Sediment Problem: Demonstrating Erosion with a Model

1. Turn on the overhead projector with the transparent pan of water on it. Set the cups into the water, and tell students that, in this model, the water represents a lake (or river), and the cups with soil on top represent hills.

2. Explain that the hills have the same amount of soil, but one is plain soil and the other has grass mixed into it.

3. Show students one tablespoon of water (or a graduated cylinder with 15 ml of water in it). Ask for predictions about what will happen if this much water "rains" on each of the hills.

4. Dribble one tablespoon of water on each hill. Ask students what they observed. [maybe little or nothing, or perhaps, "the dirt soaked up the water"] Let them know that in the real world, water that soaks into and slowly travels through the ground is called *ground water.*

5. Ask for predictions about what will happen if you "rain" another tablespoon of water on each hill. Dribble the water again, and ask for observations. If necessary, continue to dribble water, one tablespoon at a time, "raining" the same amount in the same manner on both hills.

6. Tell them that in the real world, when it rains a lot, the ground gets saturated. It cannot soak up more water, so water runs off the surface, carrying dirt with it. This is **erosion.** The dirt deposited in the water is called **sediment.**

7. If students notice that the hill with grass clippings has less sediment in the water around it, ask them why they think this is the case. [The grass "holds onto" the soil, preventing it from washing away as easily.] Point out that—in the real world—leaves and roots of plants help prevent erosion.

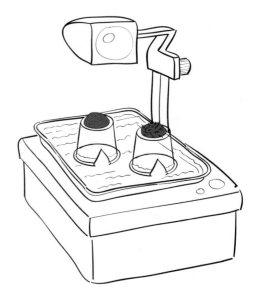

*You may want to demonstrate the idea of saturating soil and subsequent run-off using a different model; a sponge and water.*

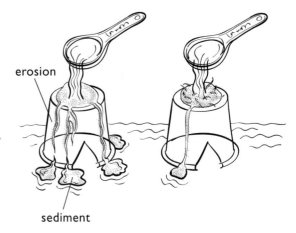

*The dirt should begin washing off the cups and clouding up the water in the tray. The water near the plain soil hill should be cloudier than near the hill with "vegetation."*

## Juan Tunó and Sedimentation in the Gray Area

1. Now bring the attention of the class back to the environmental issues in the Gray Area. Say that Juan Tunó has noticed a lot of dirt in the water in some parts of the Gray Area. Read Juan Tunó's new statement then tape it under his picture and previous statements.

### Juan Tunó
### Sediment Statement

Some of these rivers look too dirty to me. I'm not sure if too much dirt in the water might harm fish, but it seems like it would. I do know that logging, cutting down trees for lumber, can cause erosion. When trees are cut down, they don't hold down the soil anymore and the soil gets washed into streams, especially if workers log close to streams. To me, clear-cut logging is the worst, because that means they cut down every tree in an area.

2. Show them the picture of Anton Alogue, ask a student volunteer to read his statement, then tape it on the Suspect Chart.

### Anton Alogue
### Sediment Statement

You know, you guys all buy things made out of wood and paper, but you blame us for everything. We cut down trees because people need wood and paper. If you didn't buy the stuff, we wouldn't cut down the trees. We replant the forests afterwards, and we don't kill fish!

3. You may want to remind the class that the logging operations in the area are owned by Tunó Enterprises.

4. Ask the students how they think too much dirt in the water might harm fish and other water life.

## Introducing the Sediment Stations

1. Point out that some erosion of sediments into streams and lakes is of course always taking place in nature, but too much erosion can cause problems in the water. Explain that they will now conduct tests to see if any bodies of water in the Gray Area have too much sediment.

2. Show students where the learning stations are located. Briefly go over the Sediment Test Procedure Sheets. Tell them that when they have carefully completed each test, they record their results on the Test Results data sheet in their Environmental Detective Notebooks, then move on with their partner to any other available station.

3. Caution students to remove the Secchi disks after each test and place them on the paper towel.

4. Tell student pairs that when they have carefully completed all the tests, they can explore the statements and map at Headquarters and/or work with their detective notebooks.

5. You may choose to assign each pair to their first station to make sure they're evenly distributed, then have them begin.

## Small Group Discussion

1. When all pairs have finished the stations, regain the attention of the class. Tell them that each group of four will send one member to get a Sediment File.

2. Tell students there aren't too many documents in these files, so this time everyone in their group should pass them around and read the entire file.

3. Let them know that each student will also get a Sediments: What Do You Think? sheet. Everyone in the group should discuss all of the questions, and each should fill out their own sheet.

*If your students no longer need a whole-class introduction to the research and small group discussion process, you may want to have groups of four go ahead and begin their research after they finish the tests.*

Students may be surprised that there is no cloudiness or rise in temperature in the Rafta River, which flows fairly near a clear-cut area. If so, this is a good opportunity to point out that the results of scientific tests are sometimes not what we expect. Emphasize that clear cutting trees is indeed a frequent cause of erosion and sedimentation. Something about the terrain in this particular case seems to have prevented a problem. If students have noticed it in the file, discuss Anton Alogue's proposal for **more** clear-cut logging right next to the Rafta River. Do students think this is a good idea?

4. Ask one student from each group to get a Sediment File. Have the groups begin while you distribute the Sediments: What Do You Think? sheets. Circulate and help any groups having difficulty.

## Large Group Discussion

1. When the class is ready, use the questions on the Sediments: What Do You Think? sheet to facilitate a class discussion. Ask probing questions about the reasons for students' ideas to help stimulate discussion and debate.

2. Ask, "Do any of the bodies of water you tested have too much sediment?" [They should have found all the bodies of water to be fine, except for the Missterssippi River, which has too much sediment.]

3. Ask students where they think the dirt might be coming from in the Gray Area and what might be causing the sedimentation? [Among other responses, students may bring up soil erosion, logging, lack of plants, waste discharge, urban run-off, and abundant bottom feeders (such as carp) that stir up bottom sediments.]

4. Ask, "How can too much sediment harm water life?" From their research with the files, students may mention that sediment:

- May be killing fish directly by damaging and clogging their gills or covering the bottoms of rivers so fish (like salmon) can't dig the holes to lay their eggs. Or sediment could be smothering fish eggs on river bottoms.

- May also be killing fish indirectly by darkening the water, causing more solar heat to be absorbed and heating up the water. Warmer water increases the need of fish for oxygen, because it raises their metabolism. At the same time, the availability of oxygen is decreased, as warm water holds less dissolved oxygen than cold water.

- May be blocking sunlight that water plants need to grow.

- May suffocate insect eggs and the newly hatched insects that fish eat.

5. At the end of the discussion, during which your students should have mentioned Parallel Park as a possible source of sediment in the Missterssippi, introduce the hunters and fishers. Show the picture of Elmo Skeeto, have a volunteer read his statement, then tape it on the Suspect Chart.

### Elmo Skeeto
#### Sediment Statement

I represent the hunters and fishers in the area. We use the Parallel Park area a lot, because we hunt for deer, ducks, and rabbits. Six years ago they let us start hunting mountain lions again. We are very disappointed that the fish are dying off, because we also enjoy fishing. It's not our fault the fish are dying, because we have not overfished. We always throw them back when they're too small.

*Note: It will probably not yet be evident to the students what possible connection there could be between hunters and fishers in Parallel Park and the sedimentation they have just found in the Missterssippi River. That's fine. Over the course of the next activity, students will discover a possible connection.*

6. Ask your students if they have any comments on Elmo Skeeto's statement. Is there anything he said they disagree with? Any clues? Tell them that in the next class session, they will do an activity that may give them more clues about Parallel Park.

## Reflect, Predict, and Vote

1. Tell your students that, as in the previous activities, they will have a few minutes to:

- Write notes in their detective notebooks.

- Write "sediments" on their map where they think the problem comes from.

- Adjust their post-its on the chart if they have changed their ideas about what is causing the fish to die.

2. Have students put their completed Sediments: What Do You Think? sheets in their Environmental Detective Notebooks.

## Going Further

1. Suggest other erosion investigations. Students could set up their own model lakes and hills and test different materials and strategies to see how well they prevent erosion. Such investigations could also be good opportunities to use controlled experimentation, and to consider other variables involving type of soil, slope, etc.

2. Your class could perform some Secchi disk tests on actual waterways near your school.

3. Encourage students to find articles relating to erosion and sedimentation in newspapers, on other media, the Internet, etc. Are there local issues in which erosion plays a part?

# Juan Tunó
## Sediment Statement

Some of these rivers look too dirty to me. I'm not sure if too much dirt in the water might harm fish, but it seems like it would. I do know that logging, cutting down trees for lumber, can cause erosion. When trees are cut down, they don't hold down the soil anymore and the soil gets washed into streams, especially if workers log close to streams. To me, clear-cut logging is the worst, because that means they cut down every tree in an area.

## Anton Alogue
### Sediment Statement

Y̲ou know, you guys all buy things made out of wood and paper, but you blame us for everything. We cut down trees because people need wood and paper. If you didn't buy the stuff, we wouldn't cut down the trees. We replant the forests afterwards, and we don't kill fish!

## Elmo Skeeto
### Sediment Statement

I represent the hunters and fishers in the area. We use the Parallel Park area a lot, because we hunt for deer, ducks, and rabbits. Six years ago they let us start hunting mountain lions again. We are very disappointed that the fish are dying off, because we also enjoy fishing. It's not our fault the fish are dying, because we have not overfished. We always throw them back when they're too small.

**1.** Watch the disk as you slowly lower it into the water.

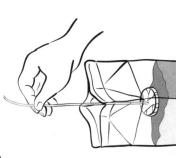

**2.** Stop lowering when you can't see the disk anymore and then pinch the string right at the surface of the water.

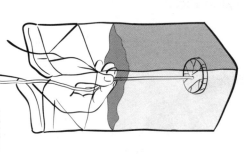

**3.** What color is the string where you pinched it?

**white** = not too much dirt in the water = good for fish

**black** = too much dirt in the water = bad for fish

**4.** Circle on your Test Results data sheet any rivers or lakes that have too much dirt for fish eggs.

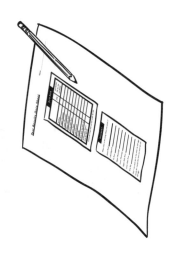

# Juan Tuno's Research Notes

How sediment pollution can affect water life:

- Sediment can cover the rocky bottoms of rivers and then fish (like salmon) can't dig the holes they need to lay their eggs
- Sediment can smother the eggs of water insects and suffocate newly hatched insect larvae
- Sediment can damage or clog fish gills so they can't breathe
- it can cut the light, making it hard for water plants to grow
- it can darken the water so it absorbs more of the sun's heat

## Juan Tuno Sediment/ Water Temperature Experiment

I left these cups with 1 cup of water in the sunlight for 1 hour. I put them all on a white tray. I wanted to see if the water that is dark with sediments will get warmer than the clear water.

| | Temperature at 10:00 AM | at 11:00 AM |
|---|---|---|
| clear water (1 cup water) | 28° C | 35° C |
| slightly cloudy water (1 cup water + 1 teaspoon dirt) | 28° C | 38° C |
| water dark with sediments (1 cup water + 2 teaspoons dirt) | 28° C | 39° C |

# Synchrony City Chronicle

## Letters to the Editor

Editor: Thanks to the clear-cut logging for this recipe. —**Name withheld by request.**

 **Recipe For Dead Fish**

- Take one river of cool water with lots of oxygen and fish.

- Add a few tons of sediment.

- Stir, then bake in the Sun until brown and warm.
  There will now be much less oxygen in the water.

- Watch those fishies float.

# Local Farmers May Be Overgrazing

*By Nicole Tucker*
EWE PEA EYE INTERNATIONAL

Local farmers are devoting more acres to grazing sheep because of high demand for wool in Synchrony City. One farmer said, "I can graze a big herd of sheep on the steep part of my land where it's hard to raise crops anyway. It's a win-win situation."

But environmentalists warn that farmers must limit grazing, especially on slopes near rivers and streams.

If sheep are allowed to overgraze, they can seriously damage the plant cover. Shrubs and grasses protect the soil and slow down water run-off that can cause erosion.

Excessive trampling by grazing animals can also make it hard for plants to grow back.

# Fish Could Be In Hot Water

*By Terry Barber*
THE AVEN NEWS

Scientists say that warming water could be killing our fish. "Warm water can't hold as much oxygen as cold water," said a local scientist.

"What's worse is that a fish's system has to work faster in warm water, so they need more oxygen. It's a double whammy—they need more oxygen, but there's less there."

## Water and Air—Have Gray Area water temperatures changed?

|  | 20 years ago | 10 years ago | 5 years ago | Today |
|---|---|---|---|---|
| Gray Bay | 58° F | 58° F | 59° F | 60° F |
| Upper Rafta River | 55° F | 55° F | 55° F | 55° F |
| Lower Rafta River | 55° F | 55° F | 55° F | 55° F |
| Lake Adaysicle | 57° F | 56° F | 57° F | 57° F |
| Upper Missterssippi | 57° F | 59° F | 58° F | 64° F |
| Lower Missterssippi | 57° F | 59° F | 58° F | 65° F |
| Fo River | 57° F | 59° F | 56° F | 60° F |
| James Pond | 58° F | 59° F | 57° F | 60° F |

© 2001 by The Regents of the University of California, LHS-GEMS. *Environmental Detectives.* **May be duplicated for classroom use.**

# Clamor About Clear Cutting

By Jaine Roston

DISASSOCIATED PRESS

Clear-cut forests and heavy winter rains have caused massive erosion, habitat destruction, and water pollution in some parts of the country. After a clear cut, the forest is gone. New trees may have trouble sprouting because of dry conditions or brush covering the ground.

In recent years, trees have been "selectively" cut in most forests. This means cutting some trees in an area, but leaving others alone. Selectively cut forests can recover more easily, and much wildlife habitat is saved.

Last year, the President signed legislation allowing logging companies to clear cut again in certain forests of unhealthy trees, called "salvage." But critics claim that some forests of healthy trees are being called "salvage," and are being clear cut.

Dear U.S. Forest Service Representative:

We would like to request that you designate the selective-cut logging area near the Rafta River as "salvage." We have already clear cut a nearby area with no harm to the environment. We think this site deserves reclassification.

Anton Alogue

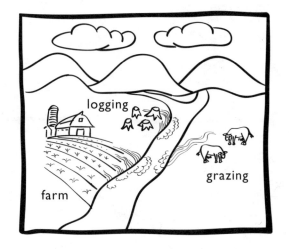

logging

farm

grazing

# Sediments: What Do You Think?

1. Do any of the rivers and lakes you tested have too much sediment? Which ones?

_____

_____

2. Where might the sediment be coming from?

_____

_____

_____

_____

3. How can too much sediment harm water life?

_____

_____

_____

_____

# Activity 5: Deer Lion

## Overview

In this activity, students play a game that simulates population dynamics. They graph a population of deer over a series of "years." Eventually, a predator (mountain lion) is introduced to the simulation, and the students graph both the deer and mountain lion populations. They then study and analyze the graphs to notice trends and draw some conclusions.

After the game, your students eagerly apply what they have learned about population dynamics to the mystery of the dying fish in the Gray Area. When did hunting mountain lions become legal in Parallel Park? Could fewer mountain lions have resulted in soaring deer populations? Did the deer consume so much vegetation that soil is being eroded into the Missterssippi River? The graphs and articles in the "Deer Lion Files" provide additional insight into these questions.

*By the end of this activity, some of your environmental detectives will understand the link between mountain lions and sedimentation without prompting, but most students will need some guidance. Ask questions such as, "How could mountain lions affect plants in the area?"* **However tempting it may be to give students the "answers," it's better to let them arrive at these important connections for themselves.**

---

### CHOOSING BETWEEN TWO VERSIONS OF THE POPULATION GAME

There are two versions of the population game, both outlined in detail on the pages that follow. Most teachers choose one version or the other, but some choose to do both. Although the game is essentially the same in both versions, it's a good idea to read the active version first, because the rules are easier to understand. Then read the dice version and decide which is best for you. Here are some key considerations in planning which version to use:

**In the active version, students stand in two lines and move back and forth to play the part of deer and resources.**

- No special materials or preparation are needed.
- The game rules are simpler to explain and less abstract.
- Populations are plotted on a large graph together as a class.
- You need a large room or space outdoors.
- Management of student behavior can be a consideration.

**In the dice version, students work in pairs at their desks with dice representing the deer and resources.**

- You will need to gather a large number of dice*.
- Students are seated.
- Students graph population changes individually.

***Some alternatives for obtaining dice are outlined on page 115.***

---

## What You Need

**For the class:**

- ❏ Headquarters set-up from previous sessions
- ❏ a chalkboard, butcher paper, or overhead transparency copy of the Population Game Graph (master on page 132)
- ❏ 2 colors of chalk, marker, or transparency pen
- ❏ 1 copy of Juan Tunó's Deer Lion statement (master on page 131)
- ❏ an overhead projector
- ❏ 1 overhead transparency of the Deer and Mountain Lion Populations in Parallel Park (master on page 134)

**For each group of four students:**

- ❏ a 9" x 12" envelope or file folder
- ❏ 1 copy of the Deer Lion Files (masters on pages 134–138)
- ❏ 1 Deer Lion Discussion Card (master on page 139)

**For each student:**

- ❏ their Environmental Detective Notebooks from previous sessions
- ❏ *(optional)* for class work or homework: 1 copy of the Deer Lion: What Do You Think? worksheet (master on page 140)

**Additional material for dice version only:**

**For the class:**

- ❏ paper cutter or scissors
- ❏ 15 small sponges (about 4 ½" x 3")
- ❏ 2 colors of permanent marker (See "Getting Ready" for guidance on colors.)
- ❏ 1 spray bottle of water for "plumping" dice
- ❏ 1 overhead transparency of the Random Attack Board (master on page 133)

**For each group of two students:**

- ❏ 20 dice (see "Getting Ready," page 115)
- ❏ 1 small plastic bag for storing dice
- ❏ 1 copy of the Population Game Graph (master on page 132)
- ❏ 1 copy of the Random Attack Board (master on page 133)
- ❏ 1 sheet of scratch paper
- ❏ 2 pens of different colors, or 1 pen and 1 pencil
- ❏ *(optional)* a shoebox in which to toss dice

## Getting Ready (for the Active Version)

1. Read pages 118–123 to make sure you understand the rules.

2. Choose a space where your whole class can form two parallel lines. Plan how to rearrange furniture, schedule the use of a multi-purpose room, cafeteria, or outside location.

3. Make the chalkboard, butcher paper, or overhead version of the Population Game Graph (master on page 132).

4. Decide where you'll display the large Population Game Graph so that it will be visible to students during the game.

## Getting Ready (for the Dice Version)

### Before the Day of the Activity

1. Read over the explanation of the dice game, pages 124–129.

2. Preparing sponge dice. If possible, ask an adult volunteer(s) to make your class set of sponge dice. The size of the cubes will depend on the thickness of your sponge. The small cellulose sponges about 3/4" thick (and usually about 4 ½" x 3") commonly sold at grocery stores work well.

    a. Use a paper cutter or scissors to cut the sponges lengthwise into "columns." The width of these columns should be the same as the thickness of the sponge.

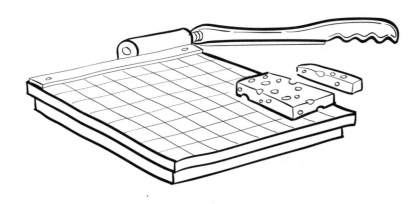

*You need 20 dice per pair of students, which is 320 dice for a class of 32 students. Having extra dice is a good idea. There are several* **alternatives** *for obtaining dice:*

- *Create a set of sponge dice. Have an adult volunteer cut up sponges to make dice, using the directions on this page. While time-consuming to make, sponge dice work well, are relatively inexpensive and durable, and they are quiet when rolled!*

- *Order small wooden or plastic cubes. Many catalogs carry cubes for mathematics activities (see "Sources," page 222). Mark two sides of each cube with the letter F, two sides with the letter S, and two sides with W (for Food, Shelter, and Water).*

- *Use regular dice. Assign two numbers to represent each resource: food, shelter, water.*

- *Use a combination of the above alternatives, with different pairs of students using different types of dice.*

*A note about color: Your dice will need to represent three things—food, water, and shelter. One easy-to-remember color code is blue = water, green = food, and yellow = shelter, but any colors practical for you will work.*

b. The fastest way to mark your dice is to color the four sides of your sponge columns before cutting them into cubes. Color two sides of each column blue and two sides green. (Or use two other colors. Don't use a marker that is the same color as your sponge.)

c. Cut the columns into cubes. The color of the sponge itself on the two cut sides will be your third color.

d. Put 20 dice in a plastic bag per group of two students. Set aside a few extra dice to replenish any that get lost.

e. Keep the permanent markers handy to refresh the colors later, since even permanent colors rub off the dice with extended use.

f. As the sponge dice dry out, the cubes become uneven. To restore them to their original shape, spray them lightly with water before class.

3. Make enough copies of the Population Game Graph (master on page 132) and Random Attack Board (master on page 133) so that there is one copy per pair of students.

4. Make the chalkboard, butcher paper, or overhead transparency version of the Population Game Graph (master on page 132).

5. Make one overhead transparency of the Random Attack Board (master on page 133).

## On the Day of the Activity

1. Plump up the sponge dice by lightly spraying them with water.

2. Have the student sheets, Juan Tunó's statement, the overheads, and the bags of dice on hand.

## Getting Ready (for the remainder of the activity)

1. Prepare the Deer Lion Files (masters on pages 134–138). Make enough copies of the articles and documents so there is one set per group of four. Put each set of articles and documents in an envelope. Label each envelope "Deer Lion Files."

2. Copy and cut apart the Deer Lion Discussion Cards (master on page 139) so there is one card per group of four.

3. Make one copy of Juan Tunó's Deer Lion statement (master on page 131).

4. Make one overhead transparency of the Deer and Mountain Lion Populations in Parallel Park (master on page 134).

5. If you've decided to use them, make one copy of the Deer Lion: What Do You Think? worksheet (master on page 140) for each student.

## Focusing on Parallel Park

1. Read Juan Tunó's Deer Lion statement to the class then tape it beneath his previous statements.

---

### Juan Tunó
#### Deer Lion Statement

I think that erosion at Parallel Park may be putting too much dirt in the Missterssippi River. I'm not sure what might be causing more erosion there now than five years ago, but I think we should take a look at the populations of some of the animals.

---

2. Ask your students what deer (and other organisms) need to survive. [food, water, shelter, and space] Accept a number of responses and discuss. Define as *resources* the things deer or other animals need to survive.

3. Define *population* as a group of one kind of animal. Ask your students, "How can a population grow?" [through many births, plenty of food and other resources] "How can populations decrease in number?" [deaths due to sickness, predators, lack of resources]

4. Ask, "How might a change in a deer population affect erosion?" [Deer eat plants. If there are too many deer, they might eat so many plants that the soil could wash away.]

5. Tell the class they're going to play a game about changes in deer populations and resources in Parallel Park.

*Note: If you're planning to play the active version of the game, see the classroom introduction that follows. Classroom steps for the dice version of the game begin on page 124.*

# The Active Version of the Population Game

water  food  shelter

## Demonstrating How to Play

1. While the class is still seated, teach them the hand signals to represent each of the resources:
      **food** (hands on stomach)
      **water** (hands cupped to mouth)
      **shelter** (hands making roof over head)

2. Ask for four volunteers to come to the front of the room to help demonstrate the rules of the game.

3. Have the four students stand facing each other in two lines. One line represents the deer population; the other line represents the resources for those deer. Mention that the whole class will form two longer lines during the real game. For the demonstration, tell them which line will represent deer, and which will represent the resources.

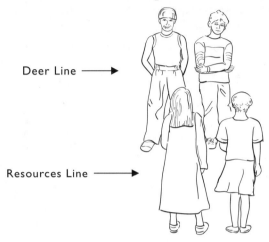

Deer Line ⟶

Resources Line ⟶

4. Show how to make a dot on the Population Game Graph to represent the number of deer you started off with in the first year (two in this demonstration). When you actually begin the game, half the number of students in your class will represent deer.

5. Have the four students turn around so the two lines are facing away from each other. Ask each student in the deer line to choose which resource (food, water, or shelter) they'll look for this "year" (round of the game). At the same time, the members of the resource line will choose which resource they will be for the year. Have each student make the appropriate hand signal. Count "1,2,3," and have the two lines turn and face each other.

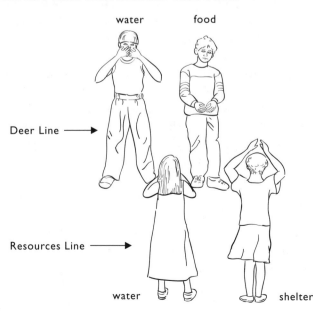

water          food

Deer Line ⟶

Resources Line ⟶

water                    shelter

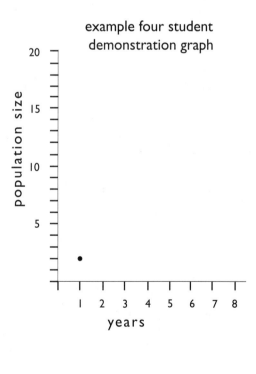

example four student demonstration graph

population size

20

15

10

5

1  2  3  4  5  6  7  8

years

6. Tell them that once they have turned, the resources will stand in place, but the deer will look for a resource with a hand signal that matches theirs. The deer then compete to gently grab onto the forearm of a resource with a matching hand signal.

- **If a deer gets a resource** (grabs onto someone holding a matching hand signal) that resource person will become a deer and move across to the deer line for the next year (round). This represents a successful deer having more offspring the following year.

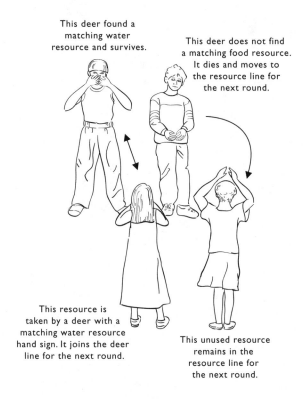

This deer found a matching water resource and survives.

This deer does not find a matching food resource. It dies and moves to the resource line for the next round.

This resource is taken by a deer with a matching water resource hand sign. It joins the deer line for the next round.

This unused resource remains in the resource line for the next round.

- **If a deer *doesn't* get a resource,** then that deer dies, decomposes, and moves across to join the resources line for the next year (round).

7. Have your volunteers go ahead and model one round.

8. Count the number of deer you have now. Did the population increase, decrease, or stay the same? Show how to make a second dot on the graph for year two, and make a line to connect the dots.

9. Choose a student to graph the class results during the real game. Let the class know that after each round, they will count the number of deer and report it to the graphing volunteer, who will record a new dot on the class graph and connect the dots.

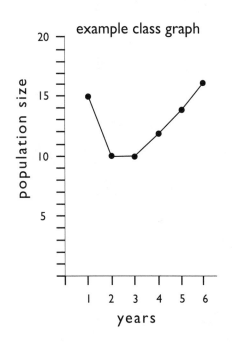

example class graph

population size

years

10. Tell students the idea is to learn how populations change in nature, so "cheating" will not benefit them in any way.

11. Emphasize that they must have their hand signal in place *before* turning around during each round (like the traditional game "roe-sham-bo"). You may want to act out the "wrong way" (turning around, scanning the opposite line, then making a hand signal that you already know matches) to help illustrate this point.

## Playing the Population Game—The Active Version

1. When your students are clear on the procedure, move the desks to clear enough space in the classroom, or move to a large room or outdoors. Display the graph, instruct your students to form two equal lines, and begin the game.

2. Once the game is underway, periodically ask your students to predict what they think will happen to the deer population in the next round by voting **thumbs up** (population will increase), **thumbs down** (population will decrease), or **thumbs horizontal** (population will remain about the same). You also may want to have the graphing volunteer trade places with someone else at some point, so they can participate in the game.

3. Continue playing for about 10 or more rounds until the "boom and crash" pattern is evident on the class graph.

*In this part of the game there are no predators, so the deer should gradually increase in number until they outnumber their resources and experience a large die-off (crash). With few deer and many resources, the deer population should then increase again. This pattern should keep repeating itself.*

*Note: If students experience a total die-off of deer, tell them another deer has "migrated" into the area, pick one student, and start off the next round with one deer.*

*Note: Some teachers prefer to stop the simulation at this point, discuss and interpret the graph, then continue the simulation with the added predator on a different day.*

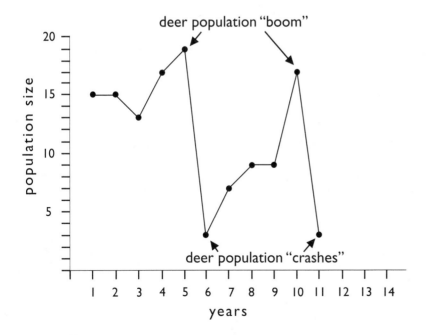

### Introducing How to Play the Game with a Predator— The Active Version

1. Tell your students they're now going to introduce a *predator*—a mountain lion—to see what impact this has on the deer population. Explain that the game will continue as before, **except** you will secretly appoint one of the members of the resource line to be a mountain lion. Tell them that the mountain lion will be "camouflaged" as a resource, so when they turn to face the resource line, they won't know which resource is actually a mountain lion in disguise.

2. The graphing volunteer will continue to record dots as before for the number of deer in each year on the graph. But from now on, the volunteer will also make a dot with a different colored marker, chalk, or transparency pen to represent mountain lions. Help the volunteer put a dot on the graph for one mountain lion in whatever year you are in at the time.

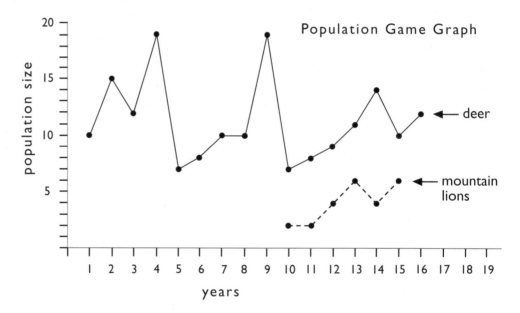

3. Let them know that just as before, any deer that do not get matched with resources die and become resources for the next round. Also as before, resources that are captured by deer become deer in the next round.

4. The difference now is if the deer grabs a mountain lion (who is disguised as a resource), then the mountain lion will roar, grab the deer back, and scare the living daylights out of them. Say that mountain lions *must* show self-restraint and patience, and not reach out and grab a deer before the deer grabs them. Once "attacked" by a mountain lion, the deer dies and joins the resource line.

5. Let the students know that from round to round the deer will never know which resources are really mountain lions in disguise, since you will choose different people each round. Tell them that only those students in the resource line whom you tap on the shoulder will be mountain lions during that round. If they were chosen as a mountain lion during a previous round they do *not* remain mountain lions—every round has a new set of mountain lions. For every mountain lion that catches a deer, there are two mountain lions hidden in the resource line in the next round. Tell the class that you will try to let everyone have at least one chance to be a mountain lion.

## Playing the Game with a Predator and Graphing— The Active Version

1. Have the students form two lines facing away from each other, with the number of deer and resources they had in the last round on the graph. Have students select and make their hand signal while you walk behind the resource line and tap one student on the shoulder to make her into a mountain lion.

2. Count "1,2,3" to begin the game again. After each round, have the graphing volunteer record the number of deer. Ask all mountain lions who *did* catch a deer during that round to raise their hands. Double that number to determine the mountain lion population for the next round, and remind the graphing volunteer to record this number as a dot on the class graph, connected with a line to the previous dot.

3. Before each round, walk behind the resource line and tap the appropriate number of students on the shoulder to make them the new population of mountain lions.

4. If a mountain lion isn't picked by a deer as a resource, it gets no meal, dies, and becomes a resource. As with the deer, if all the mountain lions are wiped out, start the next round with one mountain lion.

5. Continue to play, periodically asking the class to predict whether, in the next round, the population of deer or mountain lions will go up, down, or stay the same.

6. Once the new pattern is established on the graph, you can stop the game.

7. **Please turn to page 129, Interpreting the Graphs from *Either Version* of the Game, for the conclusion of this activity.**

*A new pattern should emerge on the graph. The deer population probably will not increase as much as before. As the deer population increases, the mountain lion population will also increase, but just behind it. As the mountain lion population increases, the deer population will decrease. The mountain lion population will then decrease, just behind the deer decrease, then the pattern should repeat itself.*

# The Dice Version of the Population Game

## Demonstrating How to Play

1. Hold up one of the dice, and explain what the colors represent: green for food, blue for water, and yellow for shelter (or whatever color code you chose). If your dice have numbers or letters instead of colored sides, explain what they represent (for example, "F" for food, "W" for water, and "S" for shelter).

2. Choose two volunteers to help you model the game. Have one partner represent the deer and the other represent the resources.

3. Give your volunteers a set of 20 dice. Have each partner start with 10—so the game will start with 10 resources and 10 deer. Each roll of the dice will represent one year.

4. Explain that when they're playing the game, one of the partners will record the deer population on their shared copy of the Population Game Graph after each round. On the class version of the graph, show where to put a dot to represent 10 deer in the first year.

5. Have both partners roll their dice and look at whatever sides are facing up.

- **The dice of the "deer partner"** represent which resources the 10 deer need this year.

- **The dice of the "resources partner"** represent the food, shelter, and water available this year.

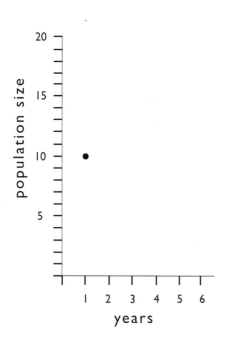

6. Show how to pair up matching dice from the two partners. For example, slide together two "shelter" dice. Have the volunteers continue until they've matched all they can. These pairs of dice will all become deer for the next round of the game. If, for example, they ended up with seven pairs of matching dice, the deer partner will have a total of 14 deer for the next year (round). This represents the successful deer having more offspring.

7. All the **unmatched** dice will go into the resources group for the next round. In our example, there should be six unmatched dice that will become resources for the next

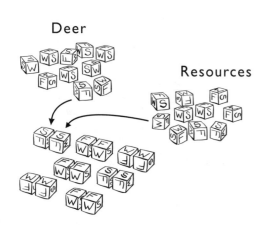

Deer

Resources

round. The three unmatched dice from the deer group represent deer that were not able to get resources, died, and became resources themselves. The three resources that were not matched represent unused resources.

14 deer

6 resources

8. On the class graph, demonstrate how to make a dot to represent the total number of deer for the next year (round), and how to draw a line connecting it to the dot from the previous year.

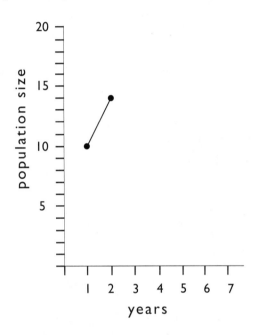

9. Tell your students that if they should experience a total die-off of deer, they should pretend another deer "migrates" into the area and start off the next round with one deer in the deer group. In other words, they will never have less than one deer.

10. Encourage students to predict to their partner, between rounds, whether they think the deer population will increase, decrease, or stay the same in the next round. Remind them to graph the deer population every round.

## Playing the Population Game—The Dice Version

1. Distribute the dice and graphs to pairs of students and let them begin. Circulate, helping students as needed.

2. Let your students continue until you see that the graphs of the majority of teams show a pattern of sharp increases and decreases in the deer population (about 10 rounds).

*In this part of the game, there are no predators, so the deer should gradually increase in number until they outnumber their resources and experience a large die-off (crash). With few deer and many resources, the deer population should then increase. This pattern should keep repeating itself.*

*If you're going to break the activity into two sessions, you may want to discuss the graphs now—before continuing with the introduction of a predator in the next session.*

3. If you plan to end the session after the class discussion of this first part of the game, collect the dice.

## Discussing the Population Game—The Dice Version

1. Have students focus on their graphs.

2. Ask for volunteers to describe what happened to the deer population in their game. If they say, for example, that the population went up and down several times, ask for anyone with similar results to raise their hands. Tell students that a sharp population increase is often called a **boom** and a major decrease is called a **crash.**

3. Facilitate a discussion, and help students interpret their graphs **for themselves,** rather than explaining them. If they miss any of the following points, guide the discussion to help bring out that:

- The deer population increased until it outnumbered its resources.

- When there weren't enough resources, lots of deer died (the deer population crashed).

*Note: This is a good place to stop if you want to break this activity into two sessions.*

- With few deer, the resources recovered and there were lots of resources. The deer population increased again, and the pattern repeated itself.

## Introducing How to Play the Game with a Predator— The Dice Version

1. Using the class graph, quickly draw a graph similar to the ones the students just completed. Draw it so that in the last round, you show 10 deer.

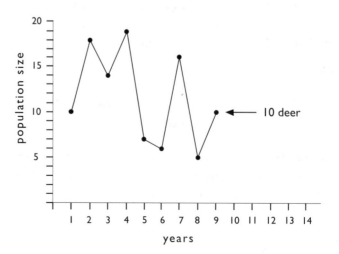

2. Tell the class they get to continue playing the game, but with the addition of a *predator*—a mountain lion—to see what impact this has on the deer population.

3. Ask for a volunteer to help you model how to play. Say you'll be the "deer partner" and the volunteer will be the "resources partner." Explain that since your sample graph has 10 deer and 10 resources in the last round, you'll take 10 dice, and have your partner take 10.

4. Put the Random Attack Board on the overhead. Tell the class that they will continue playing as before, but when they pair the dice, the deer partner will place the pairs on this board.

5. Ask the resources partner to think of a number between 1 and the total number of resources (in this case, a number between 1 and 10), and write the number on scratch paper. This secret number will be the spot on the Random Attack Board where the mountain lion is secretly hiding—the resources partner knows where the mountain lion is hiding, but the deer partner doesn't.

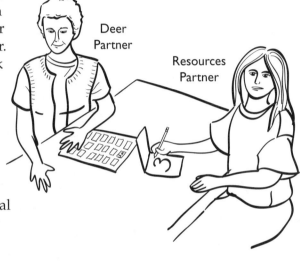

Deer Partner

Resources Partner

6. Roll your dice and have your partner do the same.

7. As the deer partner, you pair the dice, placing them on the Random Attack Board on any spot from one to the total number of resources (10 in this example). You can put them on the board in any order. Continue until all the paired dice are on the board.

8. If you happen to place a pair of dice on the mountain lion's hiding place, the resource partner can roar and turn the dice into resources. This represents a mountain lion's successful "kill." If this happens, the resource partner will have two numbers to choose as hiding places for mountain lions in the next round. (For every mountain lion that catches a deer, there are two mountain lions in the next round.)

9. If no pairs of dice are placed on the mountain lion's hiding place, that mountain lion dies, and goes back to being a resource. Tell the students that, if all their mountain lions die off, they should pretend a new one migrates into the area—so there's always at least one mountain lion hiding somewhere.

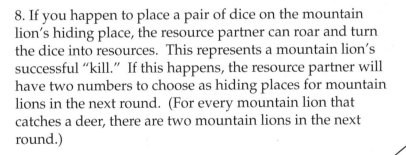

A mountain lion was secretly hiding on space #3. Both these dice die and become resources.

10. All the matched pairs of deer and resources become deer in the next round.

11. Play another sample round as a demonstration. If the mountain lion caught a deer in the last round, the resource partner will now choose **two** spaces to represent mountain lions, and write both space numbers on scratch paper. If the mountain lion didn't get a deer, the resource partner will choose one space. Remind your partner to choose a number between 1 and the total number of resources in the round.

## Playing the Game with a Predator and Graphing— The Dice Version

1. Remind partners they should start playing this next round with however many deer and resources they had in the last round they played. Their graphs can remind them of the numbers.

2. Use the class graph to show how to record mountain lions.

3. First point out the dot on the graph for the 10 deer in the last round. Use a different colored marker, chalk, or transparency pen to represent the mountain lion population that same year. (One mountain lion.)

4. Then show how to record the number of deer and mountain lions you had in the second demonstration round.

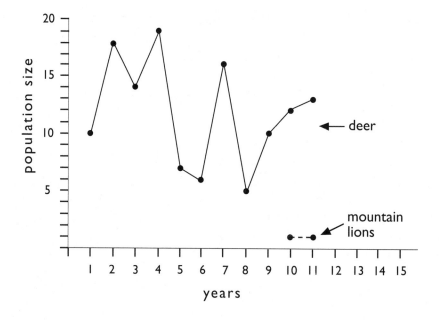

5. Distribute dice, Random Attack Boards, scratch paper, and colored pens, and have students begin. Circulate, offering help where needed. Encourage partners to switch roles as they play.

6. Once a new pattern is established on most of your students' graphs, have the class stop playing, and collect the dice.

*A new pattern should emerge on the graph. In general, as the deer population increases, the mountain lion population should also increase. At some point, there will be enough mountain lions so the deer population will decrease. The mountain lion population will then decrease, just behind the deer decrease, and then the pattern should repeat itself.*

## Interpreting the Graphs from *Either Version* of the Game

1. Review what happened in the game when there were no mountain lions. [The deer population kept getting so high that there weren't enough resources, and then the deer population dropped steeply.]

2. Ask, "What happened to the deer population when there were mountain lions?" Bring out the following points if your students do not do so themselves:

- After the mountain lions came, the number of deer didn't vary so much from year to year. The mountain lions kept the deer population from booming and then crashing.

- The mountain lion population went up and down in the same pattern as the deer, but a year or two afterward. When the deer population increased, the mountain lion population did too, and likewise when the deer population decreased so did the mountain lion population.

- When the mountain lion population was low, the deer population increased.

## Small Group Discussion

1. Ask each group of four to send one person to get a Deer Lion File and a Deer Lion Discussion Card.

2. Give groups a chance to look over the files and discuss the questions.

## Large Group Discussion

1. Put the Deer and Mountain Lion Populations in Parallel Park graph on the overhead. Ask students to interpret the graph. [The population of deer in Parallel Park went up and down steeply from 40 years ago until about 20 years ago. Then it was relatively stable until about six years ago, when it started going steeply up and down. The population of mountain lions was low until 20 years ago, then higher until about six years ago.]

2. If students don't mention it, remind them that the timeline said that hunting mountain lions started in Parallel Park around six years ago, and that the water in the nearby Missterssippi River has harmful levels of sediments.

3. Ask, "How might mountain lions affect plants in the area?" [Because of the mountain lions, the deer population wouldn't get as high, and plants wouldn't be wiped out.]

4. Ask your students how they think the simulation game relates to the mystery of the dying fish in the Gray Area. [High deer populations could be wiping out vegetation, causing erosion and sedimentation of the Missterssippi.]

## Reflect, Predict, and Vote

1. Tell your students that, as in previous activities, they will have a few minutes to:

- Write notes in their detective notebooks.

- Adjust where "sediments" is noted on their map to where they now think the problem is coming from.

- Adjust their post-its on the chart if they have changed their ideas about what is causing the fish to die.

2. If you've decided to use it, assign as homework (or have students work in class on) the Deer Lion: What Do You Think? worksheet. Have students put this sheet in their Environmental Detective Notebooks.

*Predators often help keep the population of a prey animal stable. If the predator is removed, then the prey population may begin a cycle of booms and crashes. During a boom, deer may defoliate hillsides, which causes erosion. Although in the simulation it only took one round after a crash for the resources to be plentiful again, in real life it might take a few years, or sometimes many more.*

# Juan Tunó

## Deer Lion Statement

I think that erosion at Parallel Park may be putting too much dirt in the Missterssippi River. I'm not sure what might be causing more erosion there now than five years ago, but I think we should take a look at the populations of some of the animals.

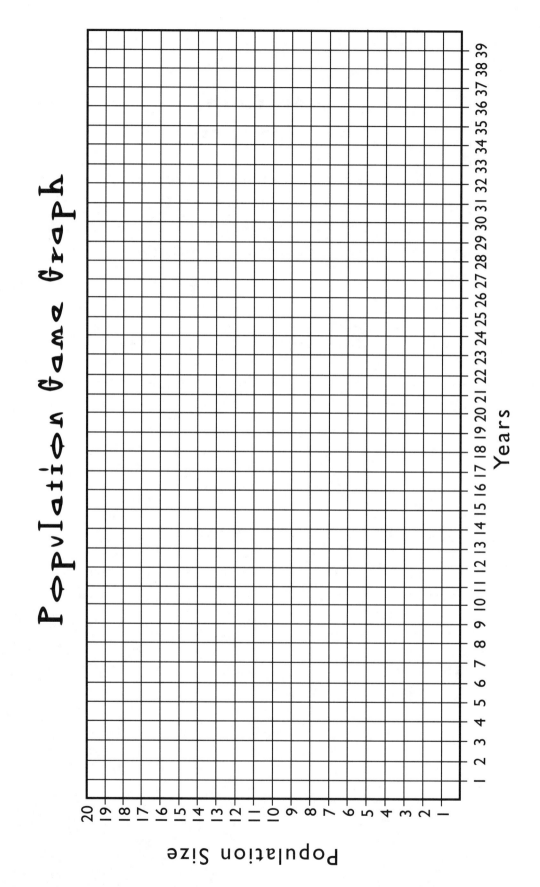

Population Game Graph

Population Size

Years
1 2 3 4 5 6 7 8 9 10 11 12 13 14 15 16 17 18 19 20 21 22 23 24 25 26 27 28 29 30 31 32 33 34 35 36 37 38 39

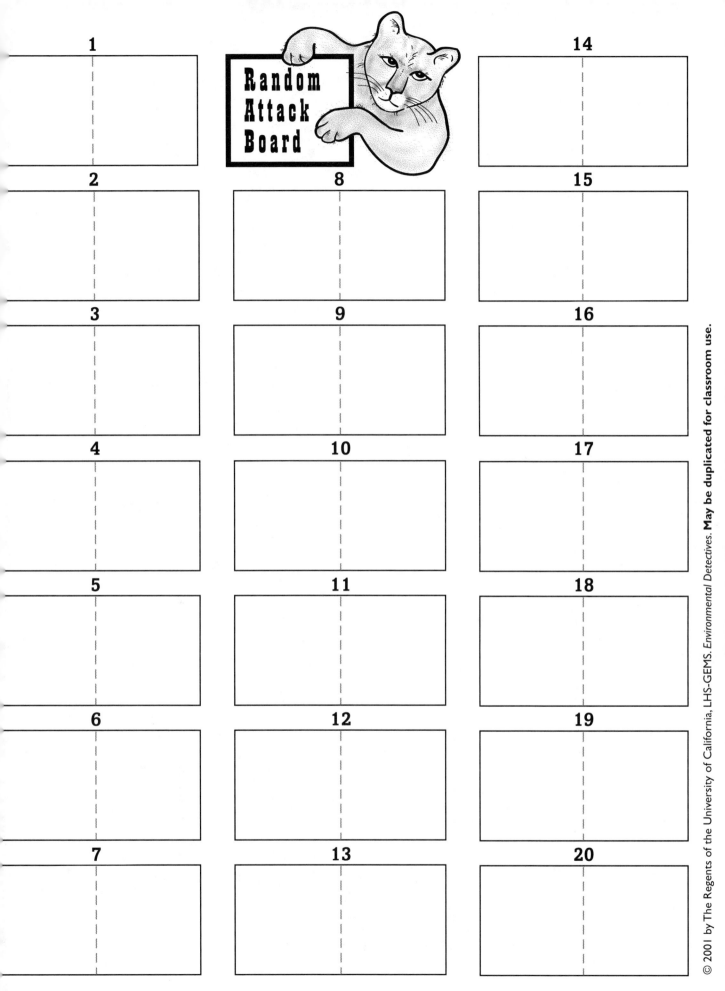

**1**

**14**

**Random Attack Board**

**2**

**8**

**15**

**3**

**9**

**16**

**4**

**10**

**17**

**5**

**11**

**18**

**6**

**12**

**19**

**7**

**13**

**20**

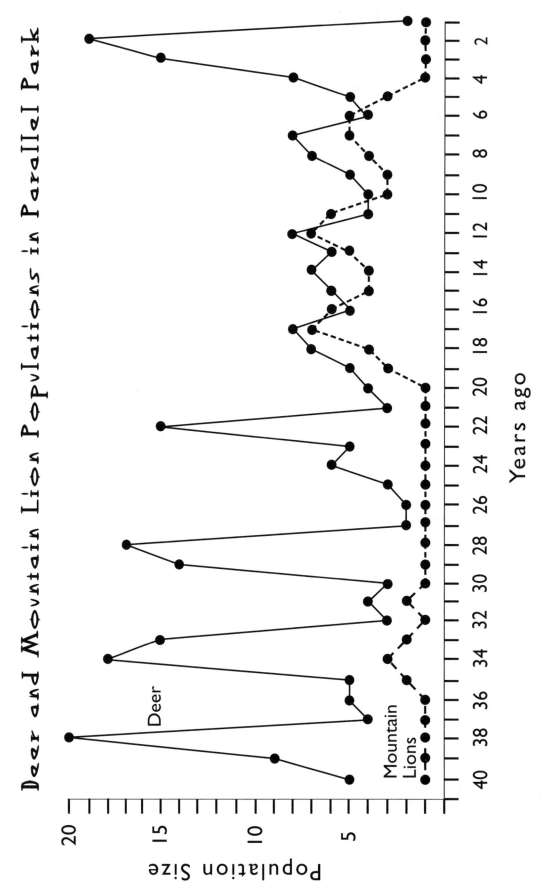

Deer and Mountain Lion Populations in Parallel Park

Population Size

Years ago

Deer

Mountain Lions

# Reference Page on Mountain Lions

## Are mountain lions dangerous?

Mountain lions, like any wild animal, can be dangerous. If you are hiking in mountain lion territory, you should talk, whistle, or sing so the mountain lions will hear you and stay away. If you meet one, try to look big. Stand up tall, remain calm, and slowly raise your arms. You can yell and throw things that you can reach without crouching down (so you stay looking tall).

## How do mountain lions help keep deer healthy?

Mountain lions eat deer. The deer they kill are usually the ones that are sick, weak, old, or young and can't get away from the mountain lions. This makes the overall population of deer healthier.

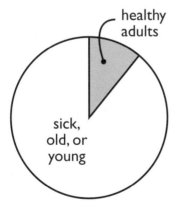

Deer Killed by
Mountain Lions
in Parallel Park

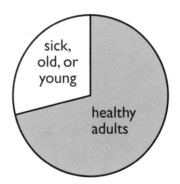

Deer Killed by
Hunters
in Parallel Park

## What do mountain lions look like?

Adult mountain lions can be up to eight feet long. They are usually tan-colored with black-tipped ears and tail and very sharp teeth.

## Popvlation of Deer in the United States

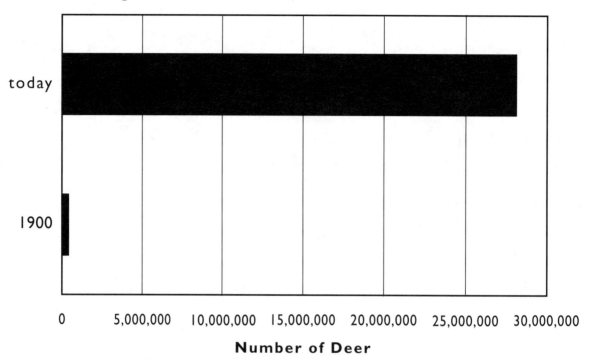

**Number of Deer**

## Plant & Deer Popvlations in Parallel Park

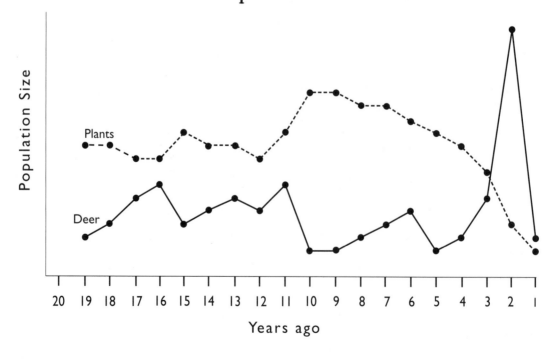

Years ago

## Mountain Biker Meets Mountain Lion

*By Matthew Milligan*
FULL COURT PRESS

Yesterday, a mountain lion lunged at a 23-year-old male mountain biker, who was riding alone on a trail.

The biker kept his bike between himself and the mountain lion until it backed off. The biker was unharmed.

Human encounters with mountain lions have become more common in areas where deer and lion hunting have been stopped. Records are sketchy, but it is likely that, in the last one hundred years, mountain lions have killed no more than about twelve people in North America.

# An Unlikely Killer

*By Karen Hosoume*
U.S. USED AND WEIRD REPORT

The most dangerous wild animal in North America may be creeping out of the woods and into your backyard. It kills hundreds of people, and injures tens of thousands. No, I'm not talking about mountain lions or bears. If you want a hint, think of Rudolf.

Today deer are responsible for far more deaths, injuries, and damage to property than any big predator on the continent.

About 150 people are killed every year in car accidents with deer, while in the last 10 years only three people have been killed by bears in the nation's parks.

Why? Because the deer population in the United States has exploded over the last one hundred years—from about 300,000 to between 18 and 28 million.

People have killed off most natural deer predators, like wolves, mountain lions, and bears, and have planted yards full of plants deer love to eat. Without predators, deer have multiplied dramatically.

Overpopulation of deer also causes damage by destroying plants in wild areas, causing erosion.

# Reference Page on Populations

**What is carrying capacity?**

Carrying capacity is the population size that an area has the resources to support. Weather, food, water, shelter, and space are usually what limit the carrying capacity of a particular habitat. Carrying capacities are usually pretty stable.

**Can carrying capacity change?**

The carrying capacity of an area can go up or down. The carrying capacity often goes up or down with the change of seasons. Sometimes other things can change the carrying capacity. For example:

Up:
If more resources become available, then the carrying capacity rises, at least for a while.
*Example:* When phosphates are added to streams through fertilizers, the carrying capacity goes up for algae, and the algae population grows like crazy.

Down:
If a resource is used up, the carrying capacity is lowered, sometimes for years.
*Example:* If too many deer eat too many plants and cause erosion of the soil, fewer plants can grow there next year.

**What are limiting factors?**

Limiting factors are the factors that keep animal and plant populations from growing out of control, by killing them or stopping them from reproducing. Examples of limiting factors include: predators, hunting, parasites, lack of resources, disease, competition, immigration, and changes in behavior and bodies that affect reproduction and survival.

**What would happen without limiting factors?**

Without limiting factors, things would change enormously! For example, one pair of houseflies could have more than five trillion descendants at the end of one year. That would be about 1000 for each person on earth. This doesn't happen because there are powerful limiting factors preventing such an increase, such as spiders, disease, pesticides, competition, rolled-up newspapers, fly swatters, etc.

## Deer Lion Discussion Card

- Look at the graph of the Parallel Park deer and mountain lion populations. What has happened to both over the last forty years in the Gray Area?
- Compare it to your timeline in your detective notebook.
- Do you think mountain lion and deer populations have anything to do with what may be killing the fish?

------------------------------------------

## Deer Lion Discussion Card

- Look at the graph of the Parallel Park deer and mountain lion populations. What has happened to both over the last forty years in the Gray Area?
- Compare it to your timeline in your detective notebook.
- Do you think mountain lion and deer populations have anything to do with what may be killing the fish?

------------------------------------------

## Deer Lion Discussion Card

- Look at the graph of the Parallel Park deer and mountain lion populations. What has happened to both over the last forty years in the Gray Area?
- Compare it to your timeline in your detective notebook.
- Do you think mountain lion and deer populations have anything to do with what may be killing the fish?

------------------------------------------

## Deer Lion Discussion Card

- Look at the graph of the Parallel Park deer and mountain lion populations. What has happened to both over the last forty years in the Gray Area?
- Compare it to your timeline in your detective notebook.
- Do you think mountain lion and deer populations have anything to do with what may be killing the fish?

# Deer Lion: What Do You Think?

1. What has happened to the deer population in Parallel Park since hunting for mountain lions was allowed six years ago? Explain.

_____

_____

_____

_____

_____

2. How could hunting mountain lions have anything to do with killing fish?

- Number the drawings in order from #1 to #5 to show the steps that can link the mountain lion hunting in Parallel Park to dead fish in the Missterssippi River.

| overgrazing of plants | fish eggs smothered by sediments | erosion of sediments | mountain lion hunting permitted | deer overpopulation |
|---|---|---|---|---|
|  |  |  |  |  |
| ____ | ____ | ____ | ____ | ____ |

- Explain each step.

_____

_____

_____

_____

_____

_____

Bijeta

I think Elmo skeeto is responsible because when he kills the mountain lions no one kills the deers and the population of the deer grew. The deer began to eat a lot of plants and grasses that makes the soil erode because then nothing is going to hold the soil. When the soil went in the water it made the water mucky that made the gills of the fishes cloged when they breathed because the water was so mucky. Anton Hogue would be responsible for the same reason too that kills the fishes.

# Activity 6: James Pond Tests

## Overview

In the first part of this activity, statements by Juan Tunó and three local characters introduce your students to another possible reason fish in the Gray Area might be dying—phosphates in James Pond. The students learn how excess phosphates can cause an explosion of algae (or *algal bloom*, pronounced AL-gul). This can in turn deplete the amount of oxygen in the water, and that could be connected to the survival of fish.

Students then conduct three tests. One of the tests is a hands-on chemistry activity—a simulated test for phosphates in the water draining from the cattle ranch, the golf course, and Gray's Land Town. The other two "tests" are paper-and-pencil simulations. In one, students check for the presence of certain aquatic organisms, and use them as indicators of the oxygen levels in the pond. In the other, they study and graph the population dynamics of birds in the area, including birds that eat fish.

Of the three James Pond tests, Birdwatching takes longest to do and includes graphing. To take this into account, the entire class, in groups of four, begins the Birdwatching activity at their seats. As work proceeds on Birdwatching, a few pairs of students at a time rotate to the learning stations for the other two tests.

In the second part of this activity, students consider the results of their investigations and the additional information found in the "James Pond Files." After a class discussion, they reflect, take notes, and may adjust their predictions about what is causing the fish to die. As an optional homework assignment they fill out a sheet on their results and ideas about James Pond.

*What happens in an algal bloom? Too many phosphates (and nitrates) build up in water. They may come from detergent, manure, or fertilizer. Phosphates are nutrients used by plants, so, when they increase, algae and other water plant populations increase. The water becomes greener and harder to see through. As the algae die, the bacteria that help decompose them increase, and the bacteria use up dissolved oxygen from the water. Fish and other water life that need plentiful dissolved oxygen will then die. Some water life such as carp that don't need as much dissolved oxygen in the water will survive until the oxygen level gets too low even for them.*

*Making some of the James Pond File research optional: This will be the last time in the unit students do research in files. Depending on your time frame and your students' research skills, you can use all or part of the files. The James Pond Files contain articles, e-mails, and clue cards. For a smaller James Pond File, include only the articles and e-mails (pages 177–181) and delete the 12 clue cards (pages 173–176). The clue cards provide deeper knowledge, but students will gather enough information without the clue cards to understand the part phosphates play in an ecosystem. Alternatively, you could give students the option of reading the clue cards for extra credit.*

# What You Need

## For the class:
- ❏ Headquarters set-up from previous sessions
- ❏ a chalkboard, butcher paper, or overhead transparency copy of the Birdwatching graph (master on the Test Results data sheet, page 79)
- ❏ 3 reaction trays with at least 3 compartments OR 3 sets of 3 plastic cups
- ❏ 1 teaspoon vinegar
- ❏ 1 stirrer (popsicle stick, coffee stirrer, or spoon)
- ❏ 3 plastic cups (8–10 oz. size)
- ❏ masking tape
- ❏ a permanent marker to write on masking tape
- ❏ 3 medicine droppers OR plastic straws cut in half to use as droppers (see "Getting Ready," page 145)
- ❏ 1 small squeeze bottle of bromothymol blue (BTB)
- ❏ access to a sink, or a dishtub with a squirt bottle of water and a sponge
- ❏ 1 copy of the What Happens in an Algal Bloom? pages (masters on pages 161–164)
- ❏ 2 copies of the Phosphate Test Procedure Sheet (master on page 165)
- ❏ 4 copies of the Water Life Procedure Sheet (master on page 166)
- ❏ 4 copies of James Pond (master on page 167)
- ❏ 4 copies of the Water Life Identification Key (master on page 168)
- ❏ 1 copy of Bo Vyne's picture and James Pond statement (master on page 159)
- ❏ 1 copy of Sandy Trapp's picture and James Pond statement (master on page 160)
- ❏ 1 copy of Juan Tunó's James Pond statement (master on page 157)
- ❏ 1 copy of Don Juan Tunó's James Pond statement (master on page 158)

## For each group of four students:
- ❏ 2 copies of the Birdwatching Procedure Sheet (master on page 169)
- ❏ 1 copy of the Bird Identification Card (master on page 170)
- ❏ 2 copies of the Birdwatching at James Pond sheets (masters on pages 171–172)
- ❏ a 9" x 12" envelope or file folder
- ❏ 1 copy of the James Pond Files (masters on pages 177–181)
- ❏ (optional) 1 set of James Pond Clue Cards (masters on pages 173–176)
- ❏ a paper clip
- ❏ 1 James Pond Discussion Card (master on page 182)

**For each student:**

❐ their Environmental Detective Notebooks from previous sessions

❐ *(optional)* for class work or homework: 1 copy of the James Pond: What Do You Think? worksheet (master on page 183)

# Getting Ready

## Before the Day of the Activity

1. Prepare materials for the Phosphate Test learning station.

   a. Label three cups: "Cattle Ranch Drainage into Pond," "Golf Course Drainage into Pond," and "Small Town Drainage into Pond."

   1.) Add about 2" of tap water to each cup.

   2.) Add 1 teaspoon vinegar to the water in the cup labeled, "Golf Course Drainage into Pond." Stir well.

   3.) Put one dropper in each cup.

*Since tap water varies, be sure to test your solutions with BTB to make sure that the Golf Course Drainage into Pond water tests yellow and the other samples test blue or green.*

*If you don't have medicine droppers, you can make droppers from drinking straws. Cut a straw in half. Take one of the pieces and bend over the top third. Pinch the doubled section and use as a dropper.*

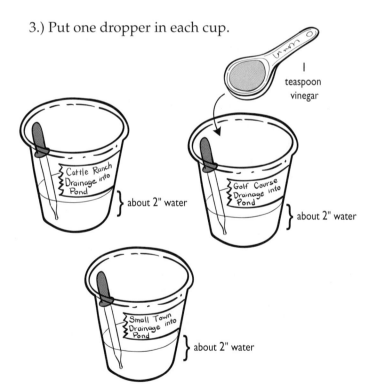

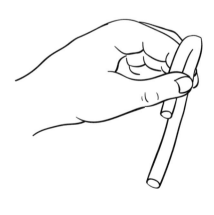

b. Prepare the "phosphate indicator." Dilute your bromothymol blue (BTB) as directed on the container it came in.

    1.) Put a small amount of BTB in the squeeze bottle.

    2.) Label the squeeze bottle, "Phosphate Indicator."

2. Make all photocopies. This activity requires a large number of photocopies. However, all are non-consumable, and can be used with multiple classes. If possible, have a volunteer make copies and organize them for you as follows:

a. for Birdwatching: Make enough copies so each group of four students has:
- 2 copies of the Birdwatching Procedure Sheet (master on page 169)
- 1 copy of the Bird Identification Card (master on page 170)
- 2 copies of the Birdwatching at James Pond sheets (masters on pages 171–172)

b. for the two learning station activities:
- 2 copies of the Phosphate Test Procedure Sheet (master on page 165)
- 4 copies of the Water Life Procedure Sheet (master on page 166)
- 4 copies of James Pond (master on page 167)
- 4 copies of the Water Life Identification Key (master on page 168)

c. for the James Pond Files:

    1.) Decide if you will include the James Pond Clue Cards in the files. If so,

        a.) Make enough copies of pages 173–176 so you have one set of 12 per group of four.

        b.) Cut up the clue cards, and clip each set with a paper clip.

        c.) Put one set in an envelope for each group.

2.) Make enough copies of the James Pond File articles and e-mails (masters on pages 177–181) so you have one set per group of four.

    a.) Put the articles and e-mails in an envelope for each group.

    b.) Label each envelope, "James Pond Files."

d. Copy and cut apart the James Pond Discussion Cards (master on page 182) so there is one card per group of four.

e. Make one copy of the four What Happens in an Algal Bloom? pages (masters on pages 161–164) for the whole class.

f. Make one copy each of the picture and James Pond statements for Bo Vyne (master on page 159) and Sandy Trapp (master on page 160), and Juan Tunó's and Don Juan Tunó's James Pond statements (masters on pages 157–158).

g. If you've decided to use them, make one copy of the James Pond: What Do You Think? worksheet (master on page 183) for each student.

## On the Day of the Activity

1. Make **one** Phosphate Test station. Set out:

- three reaction trays

- two copies of the Phosphate Test Procedure Sheet

- the squeeze bottle of phosphate indicator

- the three labeled cups of solutions you prepared earlier

- if the station is not set up next to a sink, set out a dishtub with a squirt bottle of water and a sponge

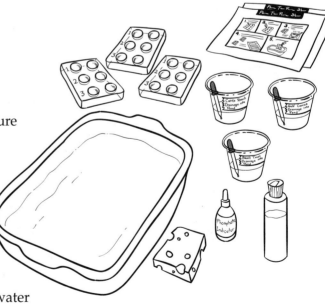

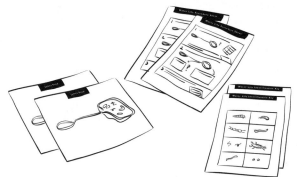

2. Set up **two** Water Life Stations: At each station, set out:

- two copies of the Water Life Procedure Sheet

- two copies of James Pond

- two copies of the Water Life Identification Key

3. Have the following Birdwatching sheets ready to distribute to each group of four:

- two copies of the Birdwatching Procedure Sheet

- one copy of the Bird Identification Card

- two copies of the Birdwatching at James Pond sheet

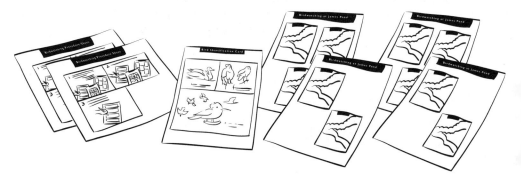

4. Post the four What Happens in an Algal Bloom? pages on the wall in a place where they can stay up for the rest of the unit, near Headquarters if possible.

5. Have on hand:

- Sandy Trapp's and Bo Vyne's pictures and James Pond statements, as well as Juan Tunó's and Don Juan Tunó's James Pond statements

- the James Pond: What Do You Think? worksheet—if you've decided to use it

6. For your introduction to the Birdwatching graph, sketch a chalkboard, butcher paper, or overhead version of the blank graph from the Test Results data sheet (master on page 79).

# Session 1: Testing James Pond

## Introducing the Phosphates Problem

1. Tell your students that Juan Tunó has some information about the pond (James Pond), read his statement, then tape it beneath his picture and previous statements.

---

### Juan Tunó
#### James Pond Statement

I think cattle manure from my uncle's ranch is killing the fish. I found a lot of phosphates in James Pond. Phosphates are a kind of chemical found in fertilizer and in manure. I know that the cattle ranch is just above the pond, and that they started keeping lots more cattle there seven years ago. I've heard that too many phosphates make too many water plants grow. Water plants make oxygen when they're alive, but when they die, they take away oxygen as they rot. I think that the fish are dying because there's not enough oxygen in the water. I think it took a few years for the phosphates to build up, then an algal bloom happened five years ago.

---

2. Point out the series of four illustrations—posted on the wall near Headquarters—representing the steps of an algal bloom. Re-read Juan Tunó's statement if necessary, or re-explain what an algal bloom is yourself. You may want to ask your students if they've ever seen an algal bloom in a fish tank or aquarium.

3. Ask for three student volunteers to read the statements of Bo Vyne (manager of the cattle ranch), Don Juan Tunó, and Sandy Trapp (golf course owner) then tape them to the suspect chart.

## Bo Vyne
### James Pond Statement

This whole area used to be farm and ranch land. A lot of that good land is under the concrete and lawns of the city. Now you want to close down our ranch just because of a test done by a kid? We don't have anything to do with fish dying. And hey, if phosphates come from fertilizer, why not look at the golf course—they put fertilizer on their pretty little putting greens, and they're always watering their lawns.

## Don Juan Tunó
### James Pond Statement

There are phosphates in *healthy* ponds! Finding phosphates doesn't prove anything! If you're looking for where they come from, TRY NATURE! Phosphates are natural! Or try Gray's Land Town! Phosphates are in laundry detergent too, ya know, and there's a lot of dirty laundry in that town!

## Sandy Trapp
### James Pond Statement

*(sarcastically)* Oh wow! Kid with chemistry set catches evil golf course in mass fish murder. That's the craziest thing I've ever heard in my life! The golf course is on the other side of the hill, and our water doesn't even drain into the pond! Besides, we already know who's really guilty. It's the killer cow patties and the dirty sock juice from the laundry!

4. Ask your students if they have any comments or responses to these statements. Ask them if they disagree with anything the suspects said. You may want to suggest that they look at their timelines (from early in the unit) to help predict who among these suspects may have caused a phosphate problem, if there is one. Let them know that they can adjust their prediction post-its on the Suspect Chart during the activity.

## Introducing the Three James Pond Tests

1. Tell the students they'll do three different tests to investigate whether phosphates are causing an algal bloom in James Pond and where the phosphates in the pond are coming from. Two are pencil and paper simulations of scientific field tests, and the other is a chemistry activity.

2. They will record the results of all three tests on the Test Results data sheet in their Environmental Detective Notebooks.

### Birdwatching

1. Explain that all groups will start with the Birdwatching test. Working as pairs in their groups of four, they will count and graph populations of different birds near James Pond. Mention that this is a simulated test, because they will look at pictures to find the birds.

2. Tell the class that, as scientists, they may be able to use the birds they find as **bioindicators.**

*If necessary, review what bioindicators are—living things used to detect something in the environment. Ask students what bioindicator was used in an earlier activity to detect chlorine in the Fo River near the water slide. [daphnia or water fleas]*

3. Hold up the Birdwatching at James Pond sheet. Briefly outline how to look at the picture of James Pond 20 years ago and count the hawks, ducks, and kingfishers. Explain that they will do the same for 15, 10, 5 years ago, and today.

4. Hold up a Bird Identification Card. Explain that the illustrations will help them identify which kind of birds they find.

*If you have time, you might want to ask students for information and experiences they may have with these and other birds.*

5. Use the sketch of the Birdwatching graph you prepared earlier to model how to put a letter "H," "D," or "K" (for hawks, ducks, or kingfishers) on the graph on the Test Results data sheet.

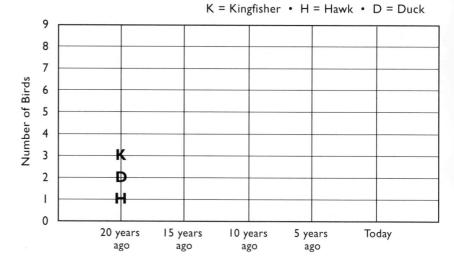

6. Tell the students that they will also have a Birdwatching Procedure Sheet to remind them of what you have just explained.

7. Explain that, while everyone is counting and graphing birds, each pair of students will get a turn to go to the learning stations to do the other tests.

**Phosphate Test**

1. Say that there is only one Phosphate Test station. At the Phosphate Test station pairs of students will test the water draining into James Pond from the cattle ranch, the golf course, and Gray's Land Town.

2. Point out that the goal is to find out if any of these drainage areas has high phosphate levels. Ask students to follow the procedure sign carefully.

**Water Life Stations**

1. Point out the two identical stations for water life, and emphasize that pairs of students should only go to one of them.

*Note: The Birdwatching Procedure Sheet includes directions on how to graph the information. For younger students, a whole-class review of how to make the line graph may be needed before the activity. On the other hand, if your students have had extensive experience with graphing, you might allow them to come up with their own system for graphing the data instead of using the method outlined on the procedure sheet at the Birdwatching station.*

*Note: The actual test for phosphates involves chemicals that are impractical for the classroom. The test the students perform is a simulated test.*

2. Tell them that if there really has been an algal bloom in James Pond and there's not much oxygen in the water, then certain small creatures that need a lot of oxygen will not be present. These creatures can be used as bioindicators to show whether there is enough oxygen in the water.

3. At the Water Life station they'll check for which kind of small creatures are present in James Pond. Briefly go over the procedure.

## Conducting the Three James Pond Tests

1. Assign a number to each pair of students. Explain that the pairs will rotate in order to the stations while everyone else is working on the Birdwatching activity. They'll go to the Phosphate Test station first. When they have recorded their phosphate test results, they'll move on to either of the two Water Life stations. When they finish the Water Life activity, they'll return to their seats and continue birdwatching and graphing. To move things along, several pairs of students can be at each station and several pairs can be circulating through the stations at once. Each station can accommodate 2 or 3 pairs of students at a time.

2. Tell them not to take materials from the stations to their desks, since the materials need to be shared with others. Remind them not to write on the materials at the stations.

3. Let them know that when they have carefully completed all three tests, their group can get a James Pond File from Headquarters so they can begin reading and discussing the contents of the file.

4. Distribute the birdwatching materials. Have students locate the Test Results data sheets in their notebooks and begin. Have the first pair of students go to the Phosphate Test station.

5. Circulate, helping students with the procedures as necessary.

6. When all of the groups have finished, collect the birdwatching materials. If you plan to break the activity here, collect any files that are in use.

## Session 2: Research and Reflection

### Small Group Research

1. Tell students they will have additional time to study the results of their three tests and do research in the James Pond Files.

2. Suggest that they skim the files and sort them into four piles. Each student can read their portion and share information with the rest of the group. If groups have improved their research skills by now, allow them to use whatever technique works best for them.

3. Say that, when they have read the files, they can get a James Pond Discussion Card and begin discussing the questions.

4. Have each group send someone to Headquarters for a James Pond File, and have them begin.

### Large Group Discussion

1. As in previous activities, facilitate a class discussion and debate.

2. For the **Birdwatching** test, ask:

- What do kingfishers, hawks, and ducks eat? [Kingfishers eat fish, hawks eat mice, and ducks eat plants and insects.]

- What does your graph tell you about fish-eating birds near James Pond? [No kingfishers are present today.]

- What does the graph show about algae-eating birds? [Ducks have increased.]

- What do these results tell you about what may be going on in the pond? [The fish population has probably declined, and the algae population has probably increased.] *Note:* Students may also propose other valid ideas about what might have happened to the fish-eating birds. For instance, Elmo Skeeto may have shot them!

---

*If you have opted for shorter files (without the clue cards) most groups will simply be able to read through the four articles and four e-mails.*

*During the discussion, the learning station test results below are likely to emerge. From these test results and their research in the James Pond Files, students should determine that phosphates in the water draining from **the golf course** have caused an algal bloom in James Pond.*

- *Birdwatching: Five years ago algae-eating birds (ducks) increased and fish-eating birds (kingfishers) disappeared. This is what the final graph should look like.*

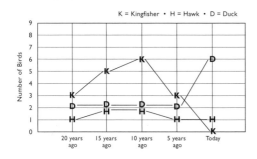

K = Kingfisher • H = Hawk • D = Duck

- *Phosphate Test:*
  —Cattle Ranch Drainage into Pond—OK phosphates
  —Golf Course Drainage into Pond—high phosphates
  —Small Town Drainage into Pond—OK phosphates

- *Water Life: James Pond is unhealthy (most of the water life that is present requires very little oxygen; water life that requires more oxygen is not present).*

3. Tell your students to look at their results from the **Water Life station.** Ask:

- What do the results at the Water Life station tell you? [Only types of water life that can survive in a low oxygen environment are now found in James Pond.]

- Judging from the data at the Birdwatching and Water Life stations, do you think that an algal bloom has happened here? Ask them if there could be any other explanation for the results.

4. Tell your students to look at their results from the **Phosphate Test.** Ask:

- Where did you find that the phosphates draining into the pond are coming from? [the golf course]

- What does a golf course use that contains phosphates? [fertilizer]

- Do you think phosphates have something to do with the fish dying? If so, how? [Phosphates act as a fertilizer, causing the algae population to increase. As algae die, they rot, using up oxygen the fish need.]

## Reflect, Predict, and Vote

1. As in the previous activity, have students:

- Write notes in their detective notebooks.

- Write "phosphates" on their map where they think the problem is coming from.

- Adjust their post-its on the chart if they have changed their ideas about what is causing the fish to die.

2. If you've decided to use it, assign as homework (or have students work in class on) the James Pond: What Do You Think? worksheet. Have students put this sheet in their Environmental Detective Notebooks.

## Going Further

**1. Create an algal bloom in a jar.** Set out two clear jars of dechlorinated water, add a cup or so of "green water" (water from a pond or fish tank) to each, and place them in direct sunlight. Add a daily pinch of fish food to one and leave the other alone as a control. Although both jars will turn greener, the rotting fish food will act as fertilizer for the algae, and the fertilized jar should turn very green within only a week or two.

**2. Calculate the exponential growth of algae.** With sunlight and fertile conditions, suppose each one-celled alga could divide in half once each night to create two algae. Have students calculate how many algae there would be in a tank in two weeks if they started with one alga.

Rajan

*mystery Conclusion*

*Which suspect(s) do you think is/are most responsible for the fish kill and why? Explain in detail.*

I think Ken Umball and Bo Vine are the main suspects responsible for the fish killing. Chlorine tests show that below the water slide, which Ken Umball owns, there are fewer water fleas than in a healthy stream, which means there's chlorine. The chlorine water can go in the oceans and hurt the fish, causing them to die and making babies deformed. Due to Ken Umball's ignorancy and stupidity the fish of the Gray's land area are dying.

Bo Vine is also a major cause of the mysterious death of the fish. She is responsible because the cow manure is scientifically proven to be bad for water. This causes the fish to die. She should have reported this to Don Juan Tuna, even though he most probably wouldn't listen; his name is proof. This uneducated twisp is also responsible for the innocent, honored fish.

# Juan Tunó
## James Pond Statement

I think cattle manure from my uncle's ranch is killing the fish. I found a lot of phosphates in James Pond. Phosphates are a kind of chemical found in fertilizer and in manure. I know that the cattle ranch is just above the pond, and that they started keeping lots more cattle there seven years ago. I've heard that too many phosphates make too many water plants grow. Water plants make oxygen when they're alive, but when they die, they take away oxygen as they rot. I think that the fish are dying because there's not enough oxygen in the water. I think it took a few years for the phosphates to build up, then an algal bloom happened five years ago.

# Don Juan Tunó
## James Pond Statement

There are phosphates in *healthy* ponds! Finding phosphates doesn't prove anything! If you're looking for where they come from, TRY NATURE! Phosphates are natural! Or try Gray's Land Town! Phosphates are in laundry detergent too, ya know, and there's a lot of dirty laundry in that town!

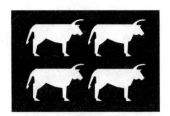

## Bo Vyne
### James Pond Statement

This whole area used to be farm and ranch land. A lot of that good land is under the concrete and lawns of the city. Now you want to close down our ranch just because of a test done by a kid? We don't have anything to do with fish dying. And hey, if phosphates come from fertilizer, why not look at the golf course—they put fertilizer on their pretty little putting greens, and they're always watering their lawns.

## Sandy Trapp
### James Pond Statement

*(sarcastically)* Oh wow! Kid with chemistry set catches evil golf course in mass fish murder. That's the craziest thing I've ever heard in my life! The golf course is on the other side of the hill, and our water doesn't even drain into the pond! Besides, we already know who's really guilty. It's the killer cow patties and the dirty sock juice from the laundry!

# What happens in an algal bloom?

**1.** Too many phosphates (and nitrates) build up in water. They may come from laundry detergent, manure, or fertilizer.

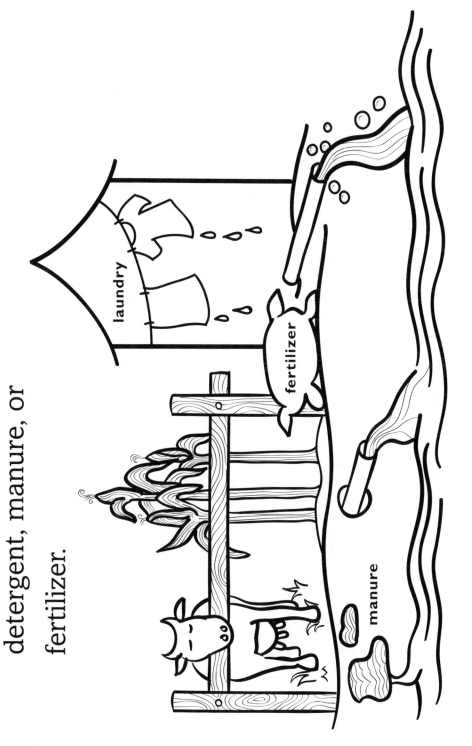

laundry

fertilizer

manure

**2.** Because there are more nutrients, lots of algae (water plants) grow. The water becomes greener, and harder to see through.

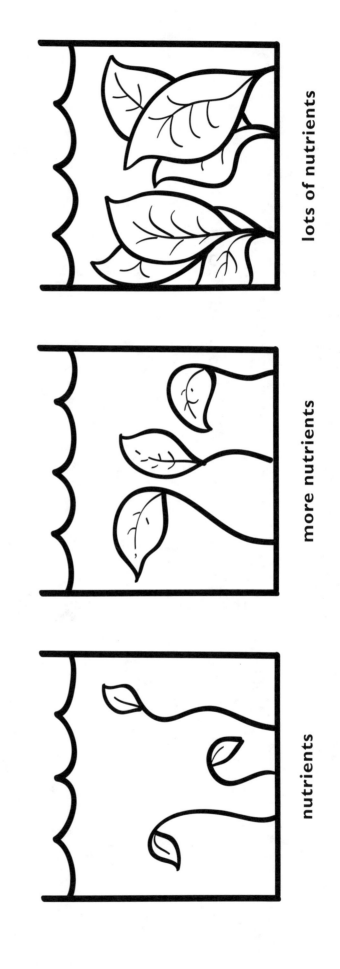

nutrients

more nutrients

lots of nutrients

**3.** As bacteria eat the dying and rotting water plants, they use up dissolved oxygen from the water.

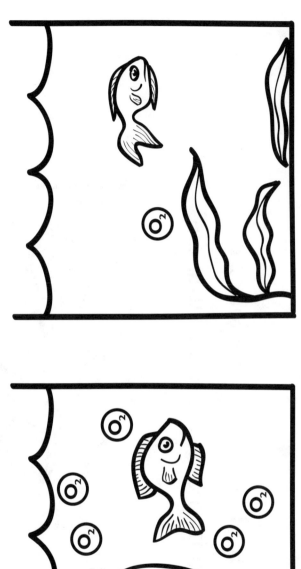

**4.** Water life that need a lot of dissolved oxygen in the water to breathe die.

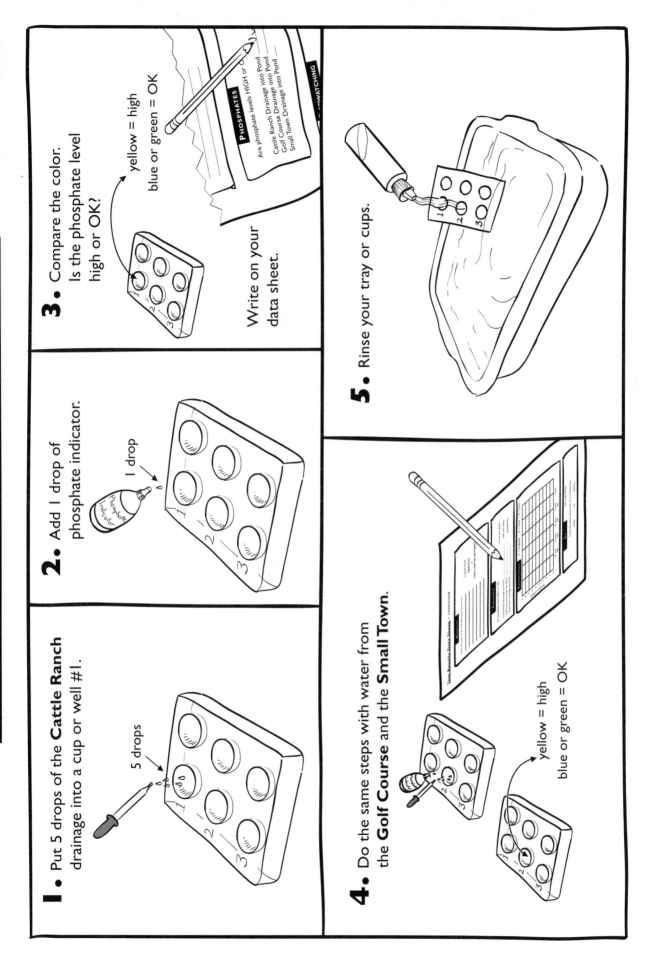

**1.** Put 5 drops of the **Cattle Ranch** drainage into a cup or well #1.

5 drops

**2.** Add 1 drop of phosphate indicator.

1 drop

Phosphate Indicator

**3.** Compare the color. Is the phosphate level high or OK?

yellow = high

blue or green = OK

PHOSPHATES
Are phosphate levels HIGH or O...
Cattle Ranch Drainage into Pond ____
Golf Course Drainage into Pond ____
Small Town Drainage into Pond ____

Write on your data sheet.

**4.** Do the same steps with water from the **Golf Course** and the **Small Town**.

yellow = high

blue or green = OK

**5.** Rinse your tray or cups.

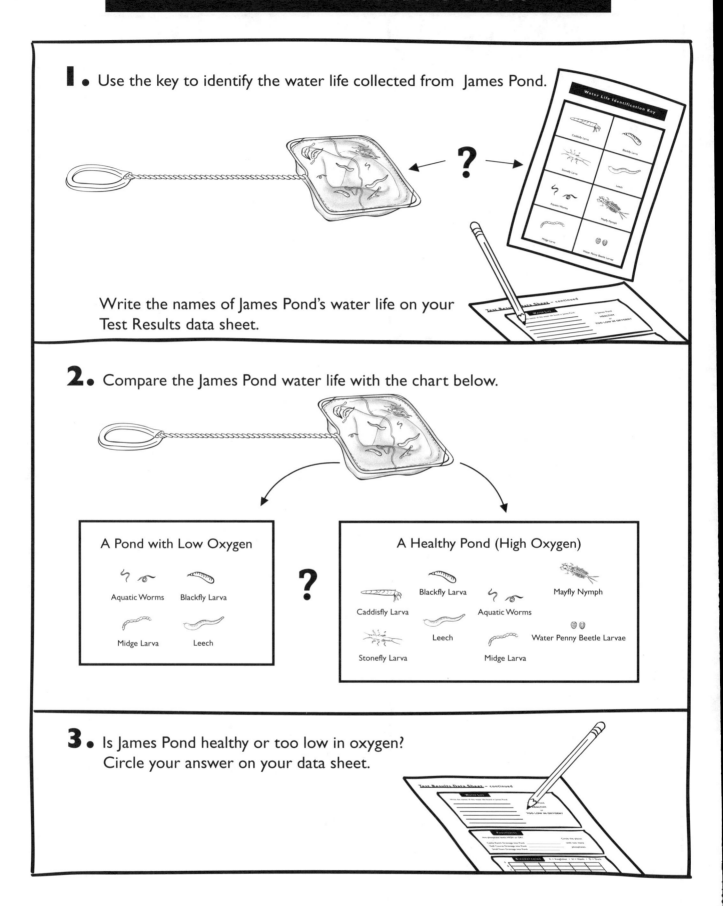

**1.** Use the key to identify the water life collected from James Pond.

Write the names of James Pond's water life on your Test Results data sheet.

**2.** Compare the James Pond water life with the chart below.

**A Pond with Low Oxygen**

Aquatic Worms   Blackfly Larva

Midge Larva   Leech

**A Healthy Pond (High Oxygen)**

Caddisfly Larva   Blackfly Larva   Aquatic Worms   Mayfly Nymph

Stonefly Larva   Leech   Midge Larva   Water Penny Beetle Larvae

**3.** Is James Pond healthy or too low in oxygen?
Circle your answer on your data sheet.

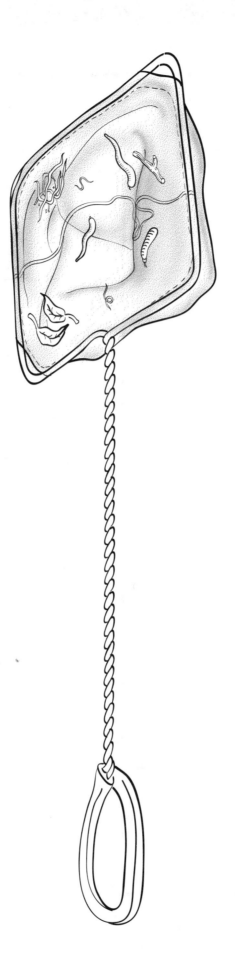

# Water Life Identification Key

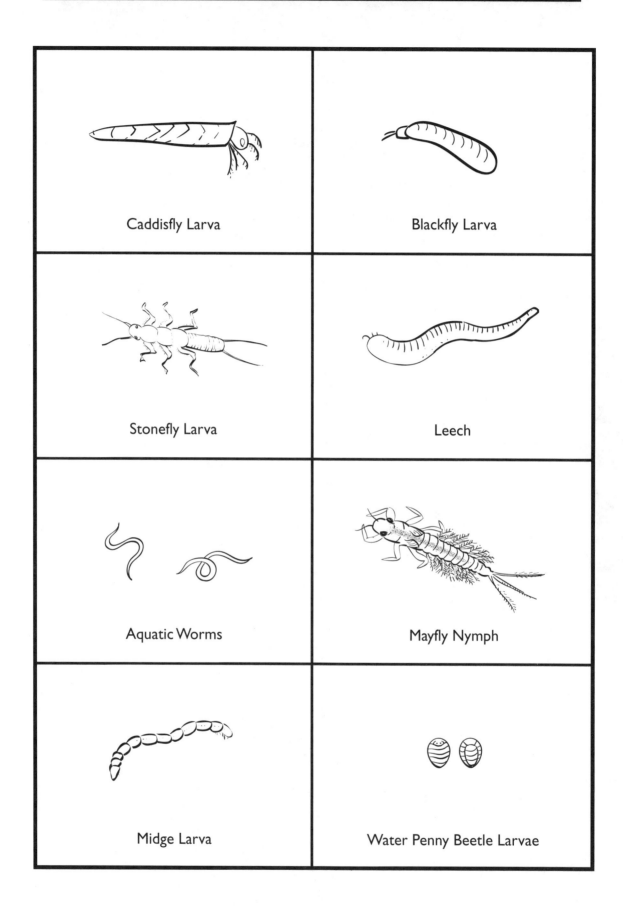

Caddisfly Larva

Blackfly Larva

Stonefly Larva

Leech

Aquatic Worms

Mayfly Nymph

Midge Larva

Water Penny Beetle Larvae

**1.** Mark on the graph how many of each kind of bird there were at James Pond 20 years ago.

Write "K" for Kingfisher
"H" for Hawk
"D" for Duck

**2.** Do the same for 15, 10, 5 years ago, and today.

**3.** Draw a line between all the Hs.
Draw a line between all the Ds.
Draw a line between all the Ks.

**Duck (Pintail)**

Eats plants and insects

**Hawk (Northern Harrier)**

/ Eats mice

**Kingfisher (Belted)**

Eats fish

20 YEARS AGO

15 YEARS AGO

# Birdwatching at James Pond

**10 YEARS AGO**

**5 YEARS AGO**

**TODAY**

## What is algae?

Algae are water plants of all sizes, in fresh water and in salt water (seaweed). All algae make food from sunshine through photosynthesis. Having some algae in ponds is important because many animals eat it. Too much algae can harm a pond.

## How does too little dissolved oxygen harm wildlife?

Many fish and water insects cannot breathe oxygen from the air. They breathe dissolved oxygen in water. Some animals (like carp, catfish, sewage worms, midge larvae, fly larvae, and leeches) need very little. Other animals (like trout, pike, salmon, and the larvae of mayflies, stoneflies, caddisflies, and beetles) need a lot. Without enough oxygen, these animals can die. Other animals that depend on them for food may die or leave.

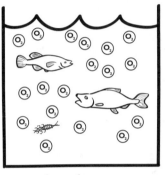

lots of oxygen          little oxygen

## What are bioindicators?

Bioindicators are living things that are sensitive to changes in their environment. Different organisms are sensitive to different types of changes. Scientists look for changes in the populations of the sensitive animals and plants to get an idea of the health of the area. Since mayflies and caddisflies will die if there is not enough oxygen in the water, they are bioindicators for lowered levels of dissolved oxygen—if they can't be found in places they usually live, it *indicates* that there may not be enough oxygen for many kinds of life.

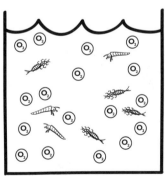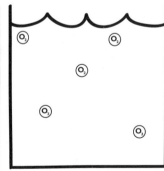

mayflies & caddisflies mean
there is oxygen

no mayflies or caddisflies
mean there is low oxygen

## What do fish breathe?

Fish do not breathe water. Fish live in water but they breathe dissolved oxygen from air bubbles that are too small for people to see. Fish tanks often have pumps that bubble oxygen, which then dissolves into the water so the fish can breathe.

JAMES POND CLUE CARD • JAMES POND CLUE CARD • JAMES POND CLUE CARD • JAMES POND CLUE CARD

## What is the geology of the area around James Pond?

The soils and rocks around James Pond are composed of limestone. Limestone is a kind of rock that is made of the same chemicals as chalk and marble (calcium carbonate). Rain water dissolves limestone easily, since rain water is usually at least a little acidic.

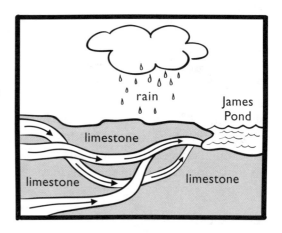

Over many years, the ground water moving through limestone can create underground rivers, caves, caverns, and sinkholes. These underground rivers can carry water much faster than it normally moves underground, and to places where you would not expect it to be by only looking at the surface of the ground.

JAMES POND CLUE CARD • JAMES POND CLUE CARD • JAMES POND CLUE CARD • JAMES POND CLUE CARD

## How do too many phosphates get in the water?

Groundwater and streams can carry phosphates into rivers and lakes if:
- animal waste is not disposed of properly, or if there are too many animals in an area
- farmers and gardeners use fertilizer and then it washes out of the soil when they water too much
- waste water from clothes and dishwashing is not treated before being released into the streams

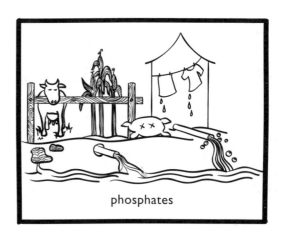

phosphates

JAMES POND CLUE CARD • JAMES POND CLUE CARD • JAMES POND CLUE CARD • JAMES POND CLUE CARD

## How are too many phosphates harmful to lakes?

Plants need phosphates to grow. Water with too many phosphates causes too many water plants (algae) to grow.

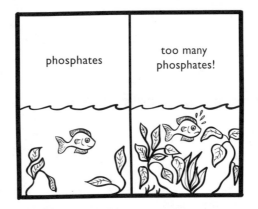

## What can cause low oxygen levels in water?

Too many water plants.
> If there are a lot of dying water plants, the bacteria eating them use up more oxygen than the living plants can make.

Warmer temperatures.
> Warm water cannot hold as much oxygen as cold water.

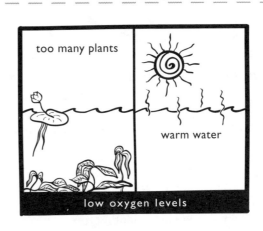

## How can too many water plants harm lakes?

Water plants make oxygen in the water. But too many water plants can cause oxygen to be taken away, because when they die and rot, the tiny animals (bacteria) that eat them may use up *all* of the dissolved oxygen in the water.

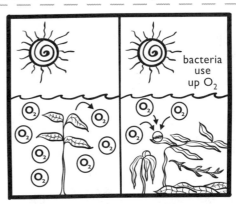

## What are phosphates?

Phosphates are chemicals that are needed for plants and animals to live and grow. There are always small amounts of phosphates in water but if water contains too many phosphates, it is unhealthy for fish. Plant growth is limited by the amount of phosphates available to them.

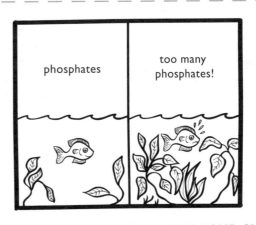

## Where does phosphate pollution come from?

Phosphate pollution comes from: human wastes, animal wastes, fertilizers, and factory wastes.

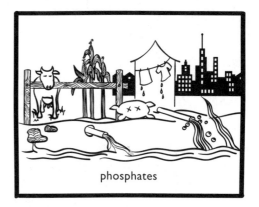

phosphates

## What happens in an algal bloom?

Too many phosphates (and nitrates) build up in water. They may come from laundry detergent, manure, or fertilizer.

Phosphates help algae (water plants) grow. The water becomes greener, and harder to see through.

As the water plants die and rot, they take away dissolved oxygen from the water.

Water life like trout that need a lot of oxygen will then die.

Some water life (such as carp) that don't need as much oxygen can live. If the oxygen level is too low, even the carp will die.

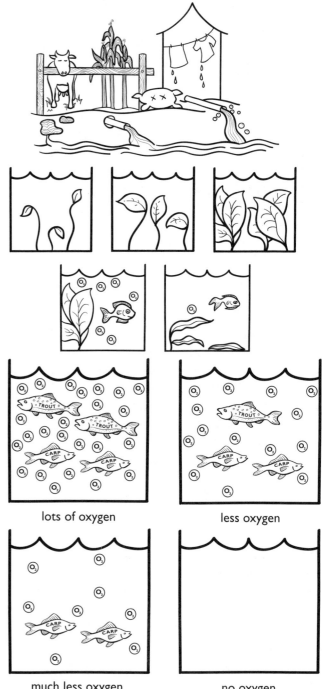

lots of oxygen

less oxygen

much less oxygen

no oxygen

# Farms Are Guilty Of Pollution—Big Time!

*By Kevin Klofkorn*
PERMANENT PRESS

Farms are the biggest polluters of rivers and streams in our country, according to the Environmental Protection Agency (EPA). They are responsible for 70 percent of waterway pollution, and foul more than 173,000 miles of waterways.

They pollute with chemicals, sediments from erosion, and animal waste. The animal waste causes algal blooms, fish kills, and non-drinkable water. The animal waste gets washed into streams from livestock areas, and from fields where it is used as fertilizer for crops.

# Lake Tohoe is Getting Murky

## Are Lawns and Cars the Culprits?

*By Laura Bergman*
BENCH PRESS

We are losing the clear water Lake Tohoe is famous for, because of algal blooms and sediment. Scientists say that cars and lawns are to blame.

The clarity of a lake is a key measure of the lake's health and is determined by how deep a Secchi disk (a white disk about the size of a din-ner plate) can be seen. In 1968, the disk could be seen in Lake Tohoe at 100 feet deep, but now it cannot be seen at 70 feet.

Algal blooms are caused by too many phosphates and nitrates in a body of water. At Lake Tohoe, the phosphates come from lawns at golf courses and homes. The nitrates come from air pollution, most of which is caused by cars in cities far away. Sediment comes from erosion, mostly from new building and construction developments.

With extra nutrients in the water, algae multiply very quickly, making the water look greener. As the algae die and rot, bacteria that eat them use up a lot of the oxygen in the water. Without much oxygen, many kinds of water life, including many fish, cannot survive.

# Petallama Dairyman In Deep Doo-Doo

*By Carl Osborn-Webb*
EWE PEA EYE INTERNATIONAL

A Petallama dairyman was fined $7,500 for polluting a nearby creek. Despite repeated warnings, he had done nothing to stop the

manure from his 350 cows from polluting a nearby creek. Every time it rained, cow manure washed into the creek. Six other dairy ranches in the area have been fined for similar problems in recent years.

Many of the dairies in the area were built next to streams long ago because the natural waterways would carry the manure away. Today, phosphates and nitrates from animal manure often cause algal blooms, which kill off entire ecosystems, including fish. Other dairy ranches in the region have spent up to $150,000 to follow pollution laws. Animal farms make 130 times the waste of the human population.

# Cormorant Kill

## Crime or Justice?

*By Lynn Baker*
KEEPA JOURNAL

Nearly 1,000 double-crested cormorants have been found shot dead on an island on one of the Great Lakes. This is the latest in a continuing controversy.

Back in the 1950s, these birds were nearly extinct because of widespread usage of the pesticide DDT, shooting/hunting, and pollution. Today the cormorant population has recovered and is strong. The island has become the largest nursery for cormorants in the country.

Cormorants eat fish, and local fishing guides say they are ruining their business. In response to the mass shootings, one fishing guide said, "I only wish they'd killed every last one." On the other hand, environmentalists called the shootings "a terrible crime" against birds that have a right to eat fish to live.

Mime-Version: 1.0
Date: Mon, 28 Sept  11:13:23
To: molawn@links
From: sandytrapp@links (Sandy Trapp)
Subject: eighth hole lawn condition

Mo
It looks like some of the grass out by the eighth hole is turning a little brown.  Let's step up the fertilizing and watering around there.

Sandy Trapp

---

Mime-Version: 1.0
Date: Mon, 28 Sept  01:20:17
To: sandytrapp@links
From: molawn@links (Mo Lawn)
Subject: re: eighth hole lawn condition

I think the browning is from some dogs that got in and peed on the grass.  It should go away by itself.  We're already using more than enough fertilizer, and watering too much, in my opinion.  I don't know where all that water goes.
Mo Lawn

---

Mime-Version: 1.0
Date: Mon, 28 Sept  3:15:05
To: molawn@links
From: sandytrapp@links (Sandy Trapp)
Subject: re: re: eighth hole lawn condition

Mo
People pay to play on green grass, that's our business.  No green grass = no business.  We need to add more fertilizer and water to our lawns.  I hope I have made myself clear.

Sandy Trapp

---

Leon—
Did you read the article about
the cormorants being killed?
Maybe we should do something
like that to the Kingfishers
around here.
          Elmo S.

## James Pond Discussion Card

- What does the information on your graph at the birdwatching station tell you?
- Do you think there has been an algal bloom in James Pond?  What makes you think so?
- If so, what do you think has caused it?

---

## James Pond Discussion Card

- What does the information on your graph at the birdwatching station tell you?
- Do you think there has been an algal bloom in James Pond?  What makes you think so?
- If so, what do you think has caused it?

---

## James Pond Discussion Card

- What does the information on your graph at the birdwatching station tell you?
- Do you think there has been an algal bloom in James Pond?  What makes you think so?
- If so, what do you think has caused it?

---

## James Pond Discussion Card

- What does the information on your graph at the birdwatching station tell you?
- Do you think there has been an algal bloom in James Pond?  What makes you think so?
- If so, what do you think has caused it?

# James Pond: What Do You Think?

1. What has happened to the kingfisher and duck populations over the last 20 years?

_____

_____

2. What do these results tell you about what may be happening to the pond?

_____

_____

_____

3. Where did you find that the phosphates draining into the pond are coming from?

_____

_____

4. Why do you think so many phosphates are coming from there?

_____

_____

_____

5. Do you think that an algal bloom has happened here and is killing the fish? Why?

_____

_____

_____

Abdullah N.

## Mystery Conclusion

I had a great time doing the mystery. I think that Don Juan Tino, Avery Wun, Ken Unball, Sandy Trapp, and Bo Vyne did the killing of the fish. I already know that the oil in the ocean matches so Avery Wun has to be a suspect. Oil kills the fish a lot! Ken Unball is also a suspect because of the chlorine in his waterslide. Don Juan Tino is a suspect too because he owns the cow ranch and that's means manure. Sandy Trapp's golf course has too many phosphates and that's not good either. I hope I get the suspects right because I am pretty sure it's them.

## Mystery Conclusion

Extra science 6/8/98

I think it is Mandy Lyfboters. She is incharge of the oil refinary. Maybe they had a huge oil spill. The oil could have gotten into the water and killed the fish. The oil and chlorine from the water slide could have killed the fish. So it could also be Ken Unball. I think it is both of them. Both of them are guilty.

---

I think that Mandy Lyfboats is guilty for killing the fish. Oil is one of the main industrys that are the couse for killing the fish. She might of spilt the oil in the river. So that is why she is guilty.

I think that Sandy Trapp is guilty for killing the fish also. The golf course is one of the main industrys for killing the fish. The golf course is very bad for the enviernment.

# Activity 7: Oil and "Who Done It?"

## Overview

Prompted by breaking news about oil found on its beaches, the Gray Area Board of Supervisors has called an emergency meeting. During the meeting, your students, along with the Board, learn about two more possible culprits in the increasingly complex mystery of the dying fish.

This time, rather than conducting their own tests and research, students find out about oil pollution through the testimony of several Gray Area characters. The class, along with the Board, then examines Juan Tunó's chromatography test data to identify what type of oil was found on the beach near Synchrony City.

*The GEMS guide* On Sandy Shores *includes an "oil on the beach" activity that examines this problem. You may want to adapt that activity, present the entire unit, or use another good resource if you and your students are interested in pursuing the problem of oil pollution.*

After the board meeting, students have an opportunity to reflect on all the information they have gathered throughout the entire unit. The class discusses which suspects are to blame for the problems raised. They then do a final vote on which of the problems they think are most responsible for the fish kill. In Activity 8, they will attempt to come up with solutions to these problems.

## What You Need

### For the class:
- ❒ Headquarters set-up from previous sessions
- ❒ 6 copies of the record of the Emergency Meeting of the Gray Area Board of Supervisors (master on pages 197–198)
- ❒ 1 copy of each of the Final Suspect Statements (masters on pages 199–201)
- ❒ an overhead projector
- ❒ 1 overhead transparency of the Oil Chromatography Test Data (master on page 195)
- ❒ 1 overhead transparency of the Fish Autopsy Results (master on page 196)
- ❒ 1 large marker
- ❒ 1 package large (2" square or larger) post-its
- ❒ 1 copy of Mandy Lyfbotes' picture and Oil statement (master on page 194)
- ❒ *(optional)* 8 name badges: Mandy Lyfbotes, Avery Wun, Don Juan Tunó, LaToya Faktorie, Anton Alogue, Elmo Skeeto, Sandy Trapp, Bo Vyne

**For each student:**

❑ their Environmental Detective Notebooks from previous sessions

## Getting Ready

1. Make six copies of the record of the Emergency Meeting of the Gray Area Board of Supervisors (master on pages 197–198) that begins this activity. Decide which students you would like to read the speaking parts. If you'd like to use them, make name badges to help identify everyone attending the meeting. If you made name badges in Activity 2, you'll already have one for the Chairperson, Juan Tunó, and Ken Unballe.

2. Make an overhead transparency of the Oil Chromatography Test Data (master on page 195). Keep a sheet of paper on hand to cover up the right side of the transparency. Also make an overhead transparency of the Fish Autopsy Results (master on page 196).

3. Make a copy of Mandy Lyfbotes' picture and Oil statement (master on page 194). Tape it on the Suspect Chart.

4. Copy and cut up the Final Suspect Statements (masters on pages 199–201) into separate slips for students to read at the end of the activity.

## Introducing the Oil Problem

1. Tell the class that an Emergency Meeting of the Gray Area Board of Supervisors has just been called. It is hoped that it will shed more light on the environmental problems.

2. Distribute the meeting records, introduce the characters, and have the selected students dramatize the meeting. If you've made them, provide name badges for each character. Encourage them to read loudly, slowly, and clearly, so everyone can take in the new information discussed at the meeting. Be prepared to display the overheads as described at the appropriate times.

3. Point out to the class that a statement for Mandy Lyfbotes is on the Suspect Chart.

# Emergency Meeting of the Gray Area Board of Supervisors

**Present:** Chairperson, Juan Tunó, Mandy Lyfbotes, Ken Unballe, Avery Wun, Don Juan Tunó, LaToya Faktorie, Anton Alogue, Elmo Skeeto, Sandy Trapp, Bo Vyne

**Speakers:** Chairperson, Juan Tunó, Ken Unballe, Mandy Lyfbotes, Avery Wun, Don Juan Tunó

**Don Juan Tunó:** I want to thank Juan and all the other scientists who have been doing tests and research. I've learned quite a bit myself, and I would like to take the opportunity to make an announcement. I'm sure the data does not point to clear cutting as the problem, but that doesn't mean it's not causing other problems. From now on, Tunó Enterprises will no longer practice clear cutting. I'd appreciate any more suggestions from Juan or other members of the audience on how to keep Tunó Enterprises from harming the environment.

**Chairperson:** Thank you Mr. Tunó, we appreciate and applaud your efforts. But unfortunately, we still haven't solved this mystery. Members of the Board, at the last meeting, some people thought chlorine from the water slide was killing the fish. Since then we have learned a lot more about other possible causes for the fish die-off.

**Ken Unballe:** Ha! I TOLD you it wasn't the water slide.

**Juan Tunó:** Mr. Unballe, we still don't know for sure what is killing the fish. And we have just discovered *another* possible reason for the fish dying—oil!

**Mandy Lyfbotes:** Excuse me, but oil cannot be the problem. I am the captain of an oil tanker. We pick up oil from where it's pumped down south, and bring it in our oil tankers to an oil refinery near Synchrony City. We have never had an oil spill anywhere near this area. Let me repeat that. WE HAVE NEVER HAD AN OIL SPILL ANYWHERE NEAR THIS AREA!

**Juan Tunó:** Oh yeah, then how do you explain that I found oil washed up on the beach near Synchrony City? Maybe you've never had a spill, but what if some of the oil tankers have illegally washed out their tanks offshore, spilling oil? And how do we know that you didn't have a small spill, but just didn't report it?

**Ken Unballe:** Yeah, oil floats on water and covers the surface. Then there is less oxygen in the water. Oil can be harmful to living things—including *(loudly)* WATER FLEAS! *(gives a stern look at Juan Tunó)*

**Avery Wun:** And FISH. I've read that oil can coat water animals and kill them. It's also poisonous if eaten. Some oil sinks to the bottom and harms the creatures that live there. It also affects the food chain. If one animal population dies from the oil, other animals that eat them will have less food to eat.

• 1 •

**Mandy Lyfbotes:** *(to Juan Tunó)* Well, "whiz kid," you're out of your league. We all know oil in the water isn't good for fish. But how do you know the oil you found is from oil tankers? Cars drip motor oil on the streets all the time. On top of that, some people change their own car oil and dump it into the gutter or into storm drains. When it rains, all this car oil drains into rivers, bays, and oceans. How do you know that's not how the oil got there?

**Don Juan Tunó:** Juan, what kind of oil *was* found washed up on the beach? Was it car oil? Refinery oil? Or was it oil from the oil tanker?

**Juan Tunó:** I'm glad you asked. I tested the oil with a gas chromatograph *(crow-mat-o-graf)* machine. When substances are put into the machine, they react with the gas inside and the machine makes different patterns. Each type of oil has its own pattern, like a fingerprint. First, here is the pattern for the oil found in Gray Bay. Transparency please?

*(Teacher puts the Oil Chromatography Test Data transparency on the overhead projector. Only the left side is revealed, showing the chromatogram of oil found in Gray Bay. The right side of the transparency—with tanker, refinery, and car oil—is covered up.)*

**Juan Tunó:** Okay, now let's see the patterns for tanker oil, refinery oil, and car oil.

*(Teacher uncovers the rest of the transparency, revealing the chromatograms for tanker oil, refinery oil, and car oil.)*

**Juan Tunó:** I think it should be obvious to everyone in this room which oil chromatogram matches the oil found in Gray Bay.

*(Teacher asks class to chime in to say which matches. They should say, "car oil.")*

**Juan Tunó:** Now that all the data is in, I think it's time to solve this mystery once and for all. The evidence shows that some of our original suspects are clearly not guilty. But who *is* guilty? I don't know about you, but I think it's more than one of the suspects in this room, and of these, one in particular is most guilty.

*(The suspects all look at each other suspiciously. They begin pointing accusing fingers at each other.)*

**Juan Tunó:** It's now time for the class out there *(points at class)* to decide who is guilty, and who is not.

• 2 •

# "Who Done It?"

1. Thank the students who participated in the role-play of the board meeting. Ask everyone to record the results of the oil chromatography test on their Test Results data sheet in their detective notebooks.

2. Announce that a news bulletin has just come in, summarizing the results of the fish autopsy report. Display the Fish Autopsy Results overhead and conduct a brief discussion of these results. Students may want to jot down these results in their notebooks.

3. Tell the class it's time to think back over the entire *Environmental Detectives* unit and arrive at a decision about why the fish are dying in the waterways of the Gray Area.

4. Ask students to name one problem that could be killing the fish. Write the name of whatever problem they mention first (chlorine, acid rain, sediments, phosphates, or oil) in bold letters with a marker on a large post-it. Ask them which of the suspects they believe is responsible for causing this problem. Ask students to back up their opinions with evidence. If there is disagreement, allow some discussion, then make a decision by voting.

5. Place the post-it above the picture of the suspect named by the students. If more than one suspect is implicated, use extra post-its.

6. Continue this process for each of the other problems and suspects they name.

7. Ask, "Which suspects don't have post-its and are probably *not guilty* in the Gray Area fish kill?" Take the pictures of these suspects off the Suspect Chart. [Because of the complexity of the mystery, classes may vary in their conclusions. When in doubt, leave the suspect on the chart.]

*Note: Depending on their developmental level and exposure to environmental issues, not all students will be ready to discuss and understand completely all of the environmental problems. That's okay. Take your students' lead. In this discussion, and in designing solutions in the next activity, tackle the problems they do identify. If you find that your students simply are not "getting" certain of the problems, you may wish to teach more in depth about those issues in the future.*

## Small Group Discussion and Vote

1. Tell the class they will now have a final small group discussion to decide which **two** of the problems (or suspects) they think are **most** responsible for the fish kill. After their discussion, each group will have two post-its to vote. Encourage them to debate and to use their data, notes, and graphs to convince each other.

2. Allow the discussions to begin and distribute two post-its to each group when you think they are ready.

3. After every group has had a chance to place their post-it votes, regain the attention of the whole class. Point out which problems/suspects ended up with the most post-its. Ask for comments. If the voting results are close, feel free to leave more than two final culprits.

4. Let students know that these are the problems they will try to find solutions for in the next session.

*If the students wish, give them a chance now to vote individually on "who" they think "done it." Ask if anyone thinks it was a combination of suspects; add these combinations to the alternatives before the vote.*

## Final Suspect Statements

1. We have provided statements for each of the suspects should your class vote that they are to blame. Recruit students to read *only* the ones their class voted for. *Note:* Regardless of the vote, the last two statements—Juan Tunó's Final statement and Avery Wun's "confession"— should be read as the closing final statements, whether your students voted for Avery Wun or not.

2. Encourage your students to respond to each suspect's statement, saying whether they agree, disagree, or want to point out a piece of evidence that supports or refutes the statement.

3. Have students read the appropriate final statements.

---

### Mandy Lyfbotes
#### Final Statement

What? Are you kidding? Your own tests proved that the oil found was CAR oil, not refinery oil! We had nothing to do with the fish kill, and you're gonna hear from my lawyers!

---

## Ken Unballe
### Final Statement

O.K., I admit that we weren't completely honest with the chlorine test, and I'm sorry for that. When we were forced to re-test on different days, we did find too much chlorine in the water. But the fish autopsies only showed chlorine in *some* of the dead fish. Something else is killing off the rest of the fish. We're too important to this community, and I will not let you shut down the water slide. The water slide is worth a few dead fish. It's the other fish killers you need to worry about!

## Don Juan Tunó
### Final Statement

I want to thank Juan and all the other scientists who have been doing tests and research, but I would like to point out that the data DO NOT POINT AT MY COMPANY. The phosphates came from the golf course, not from my cattle ranch; the oil found was from cars, not from my refinery; and the sediment was not from clear cutting. I'll do my best to keep my company from causing other problems—I just told you we won't clear cut anymore—but we didn't cause this problem.

## LaToya Faktorie
### Final Statement

Fine! Thanks a lot! We try hard to clean up our pollution, and even get an award for it, and this is how we're thanked. Yeah, sure we add to air pollution, but your own data proved that it's mostly pollution from *cars* that makes the acid rain in the Gray Area. You're probably now going to tell us we have to install scrubbers, but I'm warning you—if you do that, we'll just have to raise the price of toys to pay for it, and it'll be you who ends up paying for it in the long run!

## Anton Alogue
### Final Statement

Hey, I just do my job. I'm not the one who decides where or how we log. I need to earn a living, and logging is the only work I know. Don't take my job away, just to save some fish. Besides, the Rafta River doesn't even have a sediment problem, so we're not responsible for this fish kill.

## Elmo Skeeto
### Final Statement

There's nothing more exciting than hunting mountain lions. What a beautiful and powerful animal! It's sure complicated, but I can see how too much lion hunting has made the deer overpopulate, overgraze, and cause sediments from erosion. Maybe we should cut back a little, for the sake of the fish. I'd sure like the fishing around here to get back to normal.

## Sandy Trapp
### Final Statement

I guess you're right. Your data seems to show that phosphates from my golf course are draining into the pond and causing algal blooms. But I had no idea this was happening, and there's nothing I can do about it. That's just the way the land drains around here. You can stop those sediment and chlorine problems, but there's nothing we can do about this one.

## Bo Vyne
### Final Statement

Oh give me a break! You're just blaming us because you can't face up to closing down the golf course, the water slide, and putting a stop to mountain lion hunting. Oh yeah, and of course you can't imagine admitting your own guilt, and cut down on your driving! If you shut us down, your beef prices will go up. And let me point out one last time, phosphates and nitrates from our cattle manure are not draining into the stream or pond. We're far enough from the streams. It's the danged golf course that has caused your algal blooms!

## Juan Tunó
### Final Statement

So I guess you reached the same conclusion I did. I found that the fish were killed by a combination of a few of the suspects. But I also decided that *I* am guilty. I feel real bad about it, but as I investigated I kept realizing that I'm to blame. I use the water slide, and the evidence seems to point at it as one of the problems. I ride in cars when I could bus, bike, or walk, and the oil and acid rain evidence points to cars. The phosphates came from the golf course lawns, and I admit I've never even paid any attention to fertilizer and how much watering happens on golf courses I've been to, or even on my own family's lawn. I don't hunt, but I have contributed to erosion by riding my bike on hills off trails. It's clear I'm going to have to make some changes in what I do to help solve this problem. Not just me though, but also the suspect who I think is most responsible for the fish kill. And that suspect is—*(points index finger around the room, then finally at Avery Wun)*—Avery Wun!

## Avery Wun
### Final Statement

*(dramatically)* Okay, okay I confess! I drive a car. I feel terrible if I'm responsible for the fish dying, but I'm just like all of you! *(points at the audience)* Most of us drive or ride in cars. We all eat food raised on farms or ranches. We all use things powered by oil, use energy, wash our clothes in detergent, hunt or fish, use paper or wood products, like green lawns, go to water slides, and buy toys from the toy factory. But this is our lifestyle! It's the way we live, and there's no way to change it.

## Mandy Lyfbotes
### Oil Statement

Excuse me, but oil cannot be the problem. I am the captain of an oil tanker. We pick up oil from where it's pumped down south, and bring it in our oil tankers to an oil refinery near Synchrony City. We have never had an oil spill anywhere near this area. Let me repeat that. WE HAVE NEVER HAD AN OIL SPILL ANYWHERE NEAR THIS AREA!

# Oil Chromatography Test Data

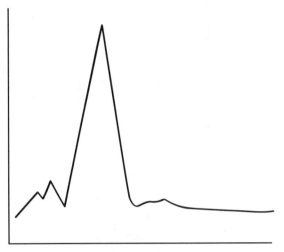

Refinery Oil

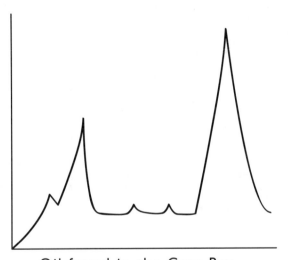

Oil found in the Gray Bay

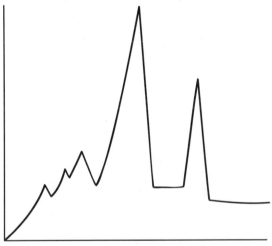

Tanker Oil

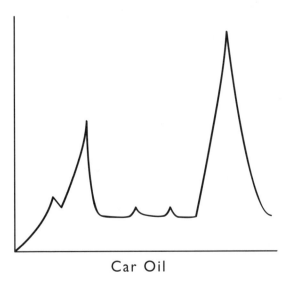

Car Oil

# NEWS BULLETIN:
# FISH AUTOPSY RESULTS

Scientists have just finished examining dead fish found in Gray Bay. Fifty fish were examined. The fish were identified as 24 Steelhead Trout, 11 Coho Salmon, 6 Carp, and 9 Northern Pike.

Some kind of oil was found in the stomachs and intestines of 14 of the fish. 21 fish had clamped fins, which indicates that they died from not enough oxygen. Chlorine was involved in the death of 8 fish, and scientists have not yet determined what killed the other 7 fish.

## Fish Dissection Results

| Oil | Insufficient $O_2$ | Chlorine | Unknown |
|-----|-----|-----|-----|
| 14 fish | 21 fish | 8 fish | 7 fish |

# EMERGENCY MEETING OF THE GRAY AREA BOARD OF SUPERVISORS

**Present:** Chairperson, Juan Tunó, Mandy Lyfbotes, Ken Unballe, Avery Wun, Don Juan Tunó, LaToya Faktorie, Anton Alogue, Elmo Skeeto, Sandy Trapp, Bo Vyne

**Speakers:** Chairperson, Juan Tunó, Ken Unballe, Mandy Lyfbotes, Avery Wun, Don Juan Tunó

**Don Juan Tunó:** I want to thank Juan and all the other scientists who have been doing tests and research. I've learned quite a bit myself, and I would like to take the opportunity to make an announcement. I'm sure the data does not point to clear cutting as the problem, but that doesn't mean it's not causing other problems. From now on, Tunó Enterprises will no longer practice clear cutting. I'd appreciate any more suggestions from Juan or other members of the audience on how to keep Tunó Enterprises from harming the environment.

**Chairperson:** Thank you Mr. Tunó, we appreciate and applaud your efforts. But unfortunately, we still haven't solved this mystery. Members of the Board, at the last meeting, some people thought chlorine from the water slide was killing the fish. Since then we have learned a lot more about other possible causes for the fish die-off.

**Ken Unballe:** Ha! I TOLD you it wasn't the water slide.

**Juan Tunó:** Mr. Unballe, we still don't know for sure what is killing the fish. And we have just discovered *another* possible reason for the fish dying—oil!

**Mandy Lyfbotes:** Excuse me, but oil cannot be the problem. I am the captain of an oil tanker. We pick up oil from where it's pumped down south, and bring it in our oil tankers to an oil refinery near Synchrony City. We have never had an oil spill anywhere near this area. Let me repeat that. WE HAVE NEVER HAD AN OIL SPILL ANYWHERE NEAR THIS AREA!

**Juan Tunó:** Oh yeah, then how do you explain that I found oil washed up on the beach near Synchrony City? Maybe you've never had a spill, but what if some of the oil tankers have illegally washed out their tanks offshore, spilling oil? And how do we know that you didn't have a small spill, but just didn't report it?

**Ken Unballe:** Yeah, oil floats on water and covers the surface. Then there is less oxygen in the water. Oil can be harmful to living things—including *(loudly)* WATER FLEAS! *(gives a stern look at Juan Tunó)*

**Avery Wun:** And FISH. I've read that oil can coat water animals and kill them. It's also poisonous if eaten. Some oil sinks to the bottom and harms the creatures that live there. It also affects the food chain. If one animal population dies from the oil, other animals that eat them will have less food to eat.

• 1 •

**Mandy Lyfbotes:** *(to Juan Tunó)* Well, "whiz kid," you're out of your league. We all know oil in the water isn't good for fish. But how do you know the oil you found is from oil tankers? Cars drip motor oil on the streets all the time. On top of that, some people change their own car oil and dump it into the gutter or into storm drains. When it rains, all this car oil drains into rivers, bays, and oceans. How do you know that's not how the oil got there?

**Don Juan Tunó:** Juan, what kind of oil *was* found washed up on the beach? Was it car oil? Refinery oil? Or was it oil from the oil tanker?

**Juan Tunó:** I'm glad you asked. I tested the oil with a gas chromatograph *(crow-mat-o-graf)* machine. When substances are put into the machine, they react with the gas inside and the machine makes different patterns. Each type of oil has its own pattern, like a fingerprint. First, here is the pattern for the oil found in Gray Bay. Transparency please?

*(Teacher puts the Oil Chromatography Test Data transparency on the overhead projector. Only the left side is revealed, showing the chromatogram of oil found in Gray Bay. The right side of the transparency—with tanker, refinery, and car oil—is covered up.)*

**Juan Tunó:** Okay, now let's see the patterns for tanker oil, refinery oil, and car oil.

*(Teacher uncovers the rest of the transparency, revealing the chromatograms for tanker oil, refinery oil, and car oil.)*

**Juan Tunó:** I think it should be obvious to everyone in this room which oil chromatogram matches the oil found in Gray Bay.

*(Teacher asks class to chime in to say which matches. They should say, "car oil.")*

**Juan Tunó:** Now that all the data is in, I think it's time to solve this mystery once and for all. The evidence shows that some of our original suspects are clearly not guilty. But who *is* guilty? I don't know about you, but I think it's more than one of the suspects in this room, and of these, one in particular is most guilty.

*(The suspects all look at each other suspiciously. They begin pointing accusing fingers at each other.)*

**Juan Tunó:** It's now time for the class out there *(points at class)* to decide who is guilty, and who is not.

• 2 •

## Mandy Lyfbotes
### Final Statement

What? Are you kidding? Your own tests proved that the oil found was CAR oil, not refinery oil! We had nothing to do with the fish kill, and you're gonna hear from my lawyers!

## Ken Unballe
### Final Statement

O.K., I admit that we weren't completely honest with the chlorine test, and I'm sorry for that. When we were forced to re-test on different days, we did find too much chlorine in the water. But the fish autopsies only showed chlorine in *some* of the dead fish. Something else is killing off the rest of the fish. We're too important to this community, and I will not let you shut down the water slide. The water slide is worth a few dead fish. It's the other fish killers you need to worry about!

## Don Juan Tunó
### Final Statement

I want to thank Juan and all the other scientists who have been doing tests and research, but I would like to point out that the data DO NOT POINT AT MY COMPANY. The phosphates came from the golf course, not from my cattle ranch; the oil found was from cars, not from my refinery; and the sediment was not from clear cutting. I'll do my best to keep my company from causing other problems—I just told you we won't clear cut anymore—but we didn't cause this problem.

## LaToya Faktorie
### Final Statement

Fine! Thanks a lot! We try hard to clean up our pollution, and even get an award for it, and this is how we're thanked. Yeah, sure we add to air pollution, but your own data proved that it's mostly pollution from *cars* that makes the acid rain in the Gray Area. You're probably now going to tell us we have to install scrubbers, but I'm warning you—if you do that, we'll just have to raise the price of toys to pay for it, and it'll be you who ends up paying for it in the long run!

## Anton Alogue
### Final Statement

Hey, I just do my job. I'm not the one who decides where or how we log. I need to earn a living, and logging is the only work I know. Don't take my job away, just to save some fish. Besides, the Rafta River doesn't even have a sediment problem, so we're not responsible for this fish kill.

## Elmo Skeeto
### Final Statement

There's nothing more exciting than hunting mountain lions. What a beautiful and powerful animal! It's sure complicated, but I can see how too much lion hunting has made the deer overpopulate, overgraze, and cause sediments from erosion. Maybe we should cut back a little, for the sake of the fish. I'd sure like the fishing around here to get back to normal.

## Sandy Trapp
### Final Statement

I guess you're right. Your data seems to show that phosphates from my golf course are draining into the pond and causing algal blooms. But I had no idea this was happening, and there's nothing I can do about it. That's just the way the land drains around here. You can stop those sediment and chlorine problems, but there's nothing we can do about this one.

## Bo Vyne
### Final Statement

Oh give me a break! You're just blaming us because you can't face up to closing down the golf course, the water slide, and putting a stop to mountain lion hunting. Oh yeah, and of course you can't imagine admitting your own guilt, and cut down on your driving! If you shut us down, your beef prices will go up. And let me point out one last time, phosphates and nitrates from our cattle manure are not draining into the stream or pond. We're far enough from the streams. It's the danged golf course that has caused your algal blooms!

## Juan Tunó
### Final Statement

So I guess you reached the same conclusion I did. I found that the fish were killed by a combination of a few of the suspects. But I also decided that *I* am guilty. I feel real bad about it, but as I investigated I kept realizing that I'm to blame. I use the water slide, and the evidence seems to point at it as one of the problems. I ride in cars when I could bus, bike, or walk, and the oil and acid rain evidence points to cars. The phosphates came from the golf course lawns, and I admit I've never even paid any attention to fertilizer and how much watering happens on golf courses I've been to, or even on my own family's lawn. I don't hunt, but I have contributed to erosion by riding my bike on hills off trails. It's clear I'm going to have to make some changes in what I do to help solve this problem. Not just me though, but also the suspect who I think is most responsible for the fish kill. And that suspect is—(*points index finger around the room, then finally at Avery Wun*)—Avery Wun!

## Avery Wun
### Final Statement

(*dramatically*) Okay, okay I confess! I drive a car. I feel terrible if I'm responsible for the fish dying, but I'm just like all of you! (*points at the audience*) Most of us drive or ride in cars. We all eat food raised on farms or ranches. We all use things powered by oil, use energy, wash our clothes in detergent, hunt or fish, use paper or wood products, like green lawns, go to water slides, and buy toys from the toy factory. But this is our lifestyle! It's the way we live, and there's no way to change it.

© 2001 by The Regents of the University of California, LHS-GEMS. *Environmental Detectives.* **May be duplicated for classroom use.**

Leah

I think that the person who was poluting the air and killing the fish was Avery Wun because she has a car. And on the two papers it shows that the car was poluting the air and water that is how the fish were dying of. It wasn't because of Don Juan Tuno or any of the other people except Avery Wun because there is not enough evidence to prove that all of the other people did it.

---

Mystery conclusion

I think the one who responsible for the fish kill is elmo sketo because he's a hunter who hunts ducks, fishes and more like all kinds of creatures. he a man who kills all kinds of animals. he enjoys to fish.

---

Mystery Conclusion

1. Which suspect(s) do you think is/are most responsible for the fish kill and why? Explain in detail.

I think that everyone is responsible except for the hunters and fishers and Mandy Lyfbotes. I think Don Juan Tuno is responsible because he owns the oil refinery which spills oil sometimes. He also owns the cow ranch which emits too many phosphates into the soil runoff. I think Avery Wun is responsible because she drives a car which also spills some oil (possibly). Ken Unbalke is responsible because he owns the water slides which, based on evidence, emit chlorine into the river below. LaToya Fakterie is responsible because she owns the toy factory which, in its smoke, produces acid rain. Anton Alogve is responsible because he owns the loggers. They cut down trees, and trees help stop erosion. When there is erosion, the soil falls into the water and may kill fish. Sandy Trappa is responsible because he owns the golf course which, through fertilizer, emits phosphates into the soil runoff. Bo Vyne is responsible because she owns the cow ranch which, through manure, also emits phosphates into soil runoff. In conclusion, I think most of the suspects are responsible for the dying of the fish.

# Activity 8: Solving the Problems

## Overview

During this final activity, students change roles from environmental detectives and scientists to become the members of the Gray Area Board of Supervisors. They brainstorm solutions to the problems they voted about during the previous activity. These solutions are written on pieces of paper, and posted under the name of the problem.

In a teacher-led class discussion, the students debate the pros and cons of each solution, and vote on whether or not to pass it. The collection of passed solutions then becomes the Board's "environmental policy." In the process of discussing the problems and their solutions, the students gain a deeper understanding of the interdependence of all parts of an ecosystem.

For a closing homework assignment, students can write to the suspect(s) implicated, explain how their pollution is killing the fish, and suggest what they could do to help solve the problem. Several "Going Furthers" for this activity are also suggested. As always, we welcome your ideas and comments!

**Special Note on "Closure."** In testing this unit, a number of teachers reported that their students were frustrated that there was not one clear "culprit" and thus no simple solution to the mystery of the fish die-off. As we note in the main introduction to this unit, this is intentional. It reflects the reality of most complex environmental problems. It also deepens, in a dramatic way, student understanding of the web of intersecting organisms, natural forces, and human impact that make up an ecosystem, within the larger environment of many interacting ecosystems. As also emphasized in the GEMS guide *Mystery Festival*, the open-ended nature of the mystery has several other positive aspects. It helps students improve their ability NOT to jump to conclusions (an important life skill if there ever was one!). It also sets the stage, all the way through and after the unit, for continuing controversy, discussion, and debate, which can be very meaningful for student learning. Finally, it symbolizes an important understanding about the nature of science—that not everything is known—the story of science is one of ongoing questioning and investigation—it does not end when a single "right answer" is found. For all of these reasons we have chosen not to provide final closure to this environmental mystery. You may want to complement this by considering a local issue of somewhat less complexity, or having your students read about real-life environmental landmarks, such as Love Canal or a major oil spill. Many literature connections and other resources are listed in the guide—we might recommend you go on from here to a wonderful book such as *Who Really Killed Cock Robin?* or another of your favorites.

## What You Need

**For the class:**

❐ Headquarters set-up from previous sessions

**For each group of four students:**

❐ 4 or more sheets of 8 ½" x 11" paper
❐ 1 large marker

**For each student:**

❐ their Environmental Detective Notebooks from previous sessions

# Small Group Brainstorming of Solutions

1. Let your students know that they are now playing the role of the Board of Supervisors, in charge of dealing with the environmental problems of the area.

2. Tell them that each group of four will choose **one** of the two problems they want to attempt to solve (for example, "chlorine").

3. When they come up with what they think is a feasible solution (for example, "make the water slide recycle their water") they should write it in bold letters with a large marker on a sheet of 8 ½" x 11" paper held horizontally, and tape it on the Suspect Chart in the column under the name of that problem.

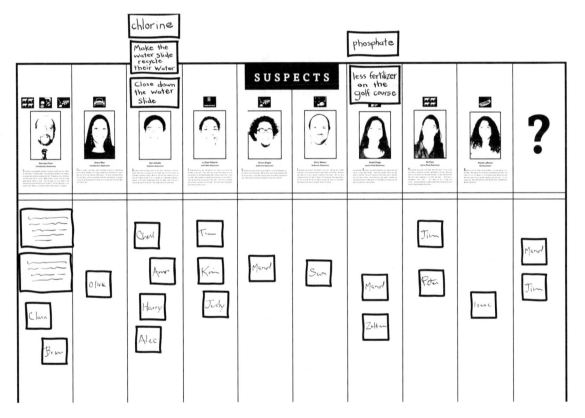

4. Remind them that they can read over their notebooks, and go to Headquarters to review the files, check suspect statements, or look at the map.

5. After taping a proposed solution under a problem column, let them know that they can then either come up with another solution to the same problem, or switch their attention to a different problem. Tell them to continue this process until you indicate to stop.

# Large Group Discussion of Solutions

1. When you have enough solution proposals posted to provide for interesting discussion, call your students' attention to the chart.

2. Focus the class on whichever problem appears to have the richest collection of solutions. Mention that it was great to get all these ideas, but now everyone should consider whether there might be a "down side" to the solutions (such as feasibility, long term vs. short term, cost vs. benefits, etc.). With the help of your students, review the problem and read one of the proposed solutions out loud.

3. Ask the class for feedback on the solution and whether or not they think it will work. Encourage discussion of the pros and cons. If students come up with more solutions at this point, have them write them on additional pieces of paper and add them to the chart.

4. When a proposed solution to the problem has been thoroughly discussed, tell the students that they may vote thumbs up or thumbs down. Any solution that has a majority in favor passes. If it has a majority against it, it does not pass. Write "Approved" on any solutions that pass.

5. Discuss and vote on all the solutions for one of the problems.

6. If time allows, begin the same process with a different problem. Do as many problem columns as you have time for and/or your students are interested in tackling. You may choose to continue on another day, or to simply end the process when you deem it appropriate.

7. When students have finished voting for solutions, explain that this is now their "environmental policy" that they have created and passed, just as a real city council would do. Let them know that as in real life, none of the solutions to the environmental problems is necessarily right or wrong. They have reached the best decision they can at this point, based on the opinions and evidence available to them. After some time has passed, they could do further tests to see if their solution is working and the fish are recovering.

# Wrapping Up the Unit

1. Tell the class Synchrony City is a play on words. The term "synchronicity" derives from the word, "synchronous," which means a number of things happening at the same time. Let them know that this name was chosen because a number of synchronous events and interactions may have caused the fish kill. This is often the case in real environmental situations, where animals, plants, people, and their physical surroundings are all **interconnected,** and where problems are often complicated.

2. Tell them "Gray Area" can mean there are few black and white answers, but a lot of "gray area." Ask them to name some specific points of "gray area" in the mystery.

3. Tell them that when facing real-life environmental, social, and political issues there are people who tend to view issues as "black and white," when in reality there is much "gray area." Even with "scientific testing" there are gray areas in data interpretation, in deciding which tests to use, and in not being aware of other possible variables.

4. Say that in their solutions they probably realized how difficult it is to come up with a "right answer," and instead were forced to come up with compromise solutions. This is very often the case in the real world. In addition, often the solution to one problem may cause or lead to others.

5. Explain that polluters are sometimes grouped as "point source" and "non-point source" polluters. "Point source" pollution comes from one point, such as a factory dumping chemicals into a stream. "Non-point source" pollutants are when pollution does not come from one particular point, but from many points, or from an area.

6. Ask them to name some point and non-point source polluters in the mystery.

7. Draw students' attention to the number of non-point source pollutants, and mention that often pollution problems are caused in part by the general public (represented by Avery Wun), which of course includes you, your students, and all of us—everyone. Point out that although many people tend to blame pollution *solely* on big companies and certain kinds of factories, this is only part of the story. Companies and large-scale industry often do create severe pollution, but it's important to understand that the general public also contributes to a great deal of pollu-

*For a historical perspective on interdependence, you may want to share the John Muir and/or Chief Sealth/Seattle quote(s):*

> *"When we try to pick out anything by itself, we find that it is bound fast by a thousand invisible cords that cannot be broken to everything in the universe."*
> *—John Muir*

> *"All things are connected."*
> *—Chief Sealth (also known as Chief Seattle), from this poetic closing to a famous speech:*

> *"How can you buy or sell the sky, the warmth of the land?*
> *Every part of this Earth is sacred.*
> *Every shining pine needle, every sandy shore;*
> *The air is precious, for all things share the same breath—*
> *The animal, the tree, the person; they share the same breath.*
> *This we know—Earth does not belong to people; people belong to the Earth.*
> *This we know—All things are connected*
> *Like the blood that unites one family—All things are connected."*

*You may choose to have your students brainstorm ways that they can use less energy, create less garbage, contribute to less air pollution, etc. in their personal lives.*

tion—so the public can be a "major contributing factor." Let students know that even point source pollution can be traced to the general public, in the sense that many factories and industries produce products that are needed, purchased, and used by the public. As has happened with a number of products, consumers can be an important force in improving the environment by letting manufacturers know they would prefer to buy the most "environmentally friendly" products.

8. In closing the unit, congratulate your students on being outstanding environmental detectives.

## Going Further

**1. The Life and Times of a Shoe.** Students take a product, such as a pair of athletic shoes, and attempt to trace the product back through all the processes and exchanges it has been through to its original source. As they do this, they should take particular note of periods or processes in the product's history where energy or other resources were used and/or where some form of pollution results. They can also map out the future of the product, again focusing on points of pollution and resource use.

**2. Is Your Local Waterway Healthy?** Your class may choose to perform some of the tests simulated in the guide or done on fictitious sources on actual waterways near your school.

# Special Assessment Task

## Letter to the Editor

On the next page is a special student assessment assignment sheet for this unit in which students read a "letter to the editor" about the fish mystery. The letter contains inaccuracies for the students to identify as they write a letter of rebuttal. A variety of alleged evidence and data is cited in the "letter to the editor" to prove its contention that Don Juan Tunó is responsible for the fish die-off.

While this assignment will demonstrate student familiarity with the environmental situation described in the guide, its intent as an assessment is much broader than that. Student responses should provide you with information on their familiarity with specific science content (the pH scale, what phosphates do in ponds, etc.) as well as their awareness that the study and solution of environmental problems is often complex. The assignment should also reflect their ability to distinguish inference from evidence and to recognize when an inference or allegation is or is not supported by evidence. We have made this an "open book" assignment to help assess their research abilities, since growth in this area is a strong goal of this guide. Of particular relevance is their ability to ferret out data and evidence that is most related to a specific topic, and the ability to marshal and communicate that evidence in support of a conclusion.

The introductory paragraph on the student assignment sheet introduces the task to them—to read and analyze the letter, then write a rebuttal letter in response. They are told that there are at least eight errors in the original letter and that their letter should cite as many of these errors as they are able to find. The student letter should also describe and explain what they think is really happening in the Gray Area, based on the data and evidence the class has gathered over the course of the unit. They are urged to be as clear and persuasive as they can.

On the page after the assignment sheet, we have included a brief summary, for the teacher only, of what we see as the errors in the letter and what the accurate data indicates. This should be helpful to you in assessing the errors students point out. Students may well come up with other inaccuracies, and/or other ways to demonstrate fallacies in the letter writer's claims. That, of course, is fine!

As with all GEMS materials, we welcome your comments on the unit and on how well this article served your purposes. We'd love to see examples of your students' responses! There are also a number of other assessment ideas summarized in the Assessment Suggestions section of this guide on page 232. And, allow us to recommend *Insights and Outcomes*, the GEMS assessment handbook, which features many effective strategies for assessing student achievement in activity-based science and mathematics, as well as case studies with a range of student work.

# Student Assignment Sheet

*Following is a letter to the editor that appeared in a Gray Area newspaper,* The Gray Times. *The letter includes at least eight errors. Your detective assignment is to find as many errors in the letter as you can, and write a letter of rebuttal pointing out these errors and explaining what your data says about what is really going on. You may use any information from your environmental detective notebook. You will be given credit for each error you find, and credit for each accurate argument or explanation from the data that you make. You will lose credit for any inaccurate data or evidence in your letter. Be as clear and persuasive as you can!*

C4 **The Gray Times** ☆☆☆☆☆

WEDNESDAY, FEBRUARY 21, 2001

Dear Editor,

Some people think chlorine from the water slide has killed the fish in our area, but this is not true! In fact, it was proven to be untrue when a chemist tested the water below the water slide and found no chlorine. So, now that we know it's not chlorine, who is to blame? Our local businessman, Don Juan Tunó, that's who! Here's why:

- His cattle ranch upstream from James Pond has leaked phosphates into the pond. Phosphates poison fish.

- Fish autopsies found oil from his refinery in the dead fish that were examined.

- Sediments from his logging operations have made the rivers dirty. The dirt in the water makes the rivers too cold for the fish.

- His oil refinery is causing pH 8 acid rain in the nearby rivers.

How's that for proof? The solution is clear and simple. Don Juan Tunó is responsible for the fish problem, and we can fix it by shutting down his operations.

Name withheld by request

# For the Teacher
## Summary of Inaccuracies in the Assessment Article

1. The chlorine chemical tests were unfairly performed a week after the last release, and the day before the next planned release of chlorinated water into the Fo River. Biological tests on water fleas (daphnia) showed chlorine is probably present.

2. Although cattle ranches often do cause excessive phosphates in adjacent waterways, the phosphates in James Pond were found to have come from the golf course, not the cattle ranch.

3. Very large quantities of phosphates could poison fish, but it's more likely that phosphates kill them indirectly by causing an algal bloom. The algae reproduce rapidly due to extra phosphate nutrients. As the algae dies, it rots, and bacteria that eat the dead plants use up oxygen. Fish die in the oxygen-depleted water.

4. Fifty fish were autopsied, and oil was found in only 14 of them. Oil found in Gray Bay was studied through gas chromatography, and found to be from car oil, not from the oil refinery or from tankers.

5. Although logging, and especially clear-cutting, often does cause sediment problems in streams, the Rafta River, which is nearest to the logging operations, does not have a sediment problem, according to Secchi disk data. The Missterssippi, which is adjacent to Parallel Park, does have too much sediment. That sediment is probably due to the killing of mountain lions, which has caused the deer to overpopulate, and eat too much vegetation, thereby causing erosion.

6. Excess sediment in the water absorbs heat, and makes the water warmer—not colder. Warm water holds less oxygen and increases fish metabolism. They need more oxygen, but there is less present. The Missterssippi River, Fo River, and James Pond have all had their water warm up in recent years.

7. The letter writer's claim about acid rain shows ignorance of the pH scale. On the scale, pH numbers from 1–6 are acidic, 7 is neutral, and 8–14 are basic. Therefore pH 8 does not indicate an acid, but a base.

8. The claim that the "oil refinery" is causing acid rain in "nearby rivers" is an inaccurate generalization. The Upper and Lower Rafta River and Lake Adaysicle *are* too acidic, with a pH of about 3. But the other waterways tested have a pH of about 5, so they are slightly acidic, but not excessively so. As regards the allegation that the refinery is solely to blame, acid rain is generally formed from pollution in the entire area, and cannot be pin-pointed to one particular source. In addition, the Gray Area acid rain is 88% nitric acid, which is primarily caused by cars and trucks, not refineries.

*Note: Your students may point out that—in general—solutions to environmental problems are not "clear and simple," as the letter writer claims, but instead involve examination of a great deal of evidence, an understanding of the interaction of many factors, and often require complex compromises and negotiations. Since this is one of the lessons of the unit, you may want to give additional credit to students who articulate this idea.*

# *Behind the Scenes*

The following information is intended to provide succinct scientific background for the teacher and may be helpful in responding to student questions. It is not meant to be read out loud to or copied for students. Given the scope of this teacher's guide, it should be evident that this section is but the tip of the iceberg. You and your students may want to do more research on these and related topics, or further investigate environmental issues in your own community. You may also want to delve into other excellent programs, such as those in the "Resources" section, starting on page 224.

## Chlorine

### What is chlorine? Is it found in nature?

Chlorine (Cl) is a chemical. It can be found just about everywhere on Earth—in rocks and minerals, the ocean, plants, animals, and the human body. It is the 11th most abundant element in the Earth's crust. The chemical symbol for chlorine is Cl. Chlorine can be made through the breaking apart of water ($H_2O$) and salt (NaCl), making chlorine (Cl), hydrogen gas ($H_2$), and sodium hydroxide (NaOH).

### Is there chlorine in salt?

Yes. Another name for salt is sodium **chloride** (NaCl), and it contains more than 60% chlorine (Cl). Any chemical with "chloride" in its name has chlorine in it.

### Can chlorine hurt people?

We use chlorine bleach to clean and brighten our clothes and bathrooms. Chlorine is also used in drinking water, pools, and hot tubs to kill bacteria that could harm people. So, in small amounts in water, chlorine can be helpful, but it can be harmful in large amounts. When chlorine is in the form of a gas it can be poisonous. Chlorine gas has been used as a weapon in World War I and the Gulf War.

### Can chlorine harm water life?

Too much chlorine in fresh water can kill many things. Tiny water life—such as daphnia (water fleas) and paramecia—are among the first life forms killed by too much chlorine. A drop in their numbers can signal that a waterway is in danger, so these organisms are often used as bioindicators.

## How can you test for chlorine?
Both chemical tests and biological tests are used.

### Chemical tests
*Pro:* Tells you for sure whether or not there is chlorine present.
*Con:* Because chlorine breaks down quickly in water, a negative test could just mean that there's no chlorine now, but there could have been earlier.

### Biological tests (like testing for water fleas)
*Pro:* Tests if there has been chlorine over a long period of time, not just that moment.
*Con:* Lack of water fleas could mean there's chlorine, but could also mean there's some other chemical, that there is not enough oxygen, or that the water is too acidic or basic. If water is too acidic or too basic, water life can't live in it.


# Acid Rain

*Note:* Please see the GEMS guide *Acid Rain* for more background information; the following is adapted from that guide. There is an excellent Environmental Protection Agency web site on acid rain at http://www.epa.gov/ebtpages/airairpoacidrain.html

### What are acids and bases?
Acids are chemicals with certain properties (like sour-tasting lemon juice and vinegar), and bases are chemicals with other properties (slippery chemicals like soap and bleach). The pH scale is used to measure how acidic or basic a chemical or substance is. On a pH chart, the numbers 0–6 represent acids; the numbers 8–14 represent bases; and pH 7 represents neutral (such as pure water). The abbreviation pH stands for "the power of hydrogen," and is a measure of the number of hydrogen ions or protons in a given volume of solution. Acids *release* hydrogen ions in a solution, and bases *remove* hydrogen ions from solution. The pH scale is a **logarithmic scale,** like the Richter scale used to measure ground movement in earthquakes. So, on the pH scale, a substance of pH 6.0 is ten times more acidic than a substance of pH 7.0, a substance of pH 5.0 is one hundred times more acidic than a substance of pH 7.0, a substance of pH 4.0 is one thousand times more acidic than a substance of pH 7.0, and so on.

## What is acid rain?  Where does it occur?

Acid rain is rain with a pH below 5.6.  Sulfuric acid and nitric acid account for 95% of the acids in acid rain.  The sulfuric acid type is most commonly formed in areas that burn coal for electricity, such as the Northeastern United States.  Acid rain formed by nitric acid is more common in areas with a lot of internal combustion engines (such as in automobiles)—the Los Angeles basin for example.  The area most affected in the U.S. has been the Northeast, where average rainfall is pH 4.0–4.5.  Storms with a pH as low as 3.0–4.0 are not unusual, and strongly acidic values under pH 3.0 have been reported.  The area in the U.S. with the most rapid increase in acid precipitation seems to be the Southeast.  West of the Mississippi River seems somewhat better, except in certain troublespots, such as the Los Angeles Basin, the San Francisco Bay Area, and parts of Colorado.  In other parts of the world rainfall has also become increasingly acidic, including Canada, Scandinavia, Germany, and the British Isles.  Acid rain is a common feature of most large cities.

## How is acid rain formed?

Acid rain is formed when certain human-made air pollutants travel up into the atmosphere and chemically react with moisture and sunlight to produce acids.  These acids dissolve in tiny droplets of water vapor in the clouds and are rained down onto the earth.  While it is true that air pollutants that contribute to the formation of acid rain are and have long been released by natural processes (such as volcanoes and the activity of soil bacteria), it has only been with the industrial revolution and the internal combustion engine that these gaseous pollutants have been produced in large enough amounts to significantly affect the acidity of rain.  Oxides of sulfur and nitrogen are the main gases emitted into the air, mostly from fossil fuel combustion in power plants, industry, and from cars and trucks.  Oxides of sulfur (such as $SO$, $SO_2$, and $SO_3$) come mainly from coal burning industry and power plants.  $SO_x$ is shorthand for describing the mixture of these three gases.  Oxides of nitrogen (such as $NO$ and $NO_2$) come mainly from cars and trucks.  $NO_x$ is shorthand for describing the mixture of these gases.

## How is acid rain harmful?

There are three major harmful effects of acid rain on an ecosystem: through contact with plants, contact with soil and water, and movement of trace metals.  These effects on plants, soil, and water in turn affect other living organisms that depend on them. Acid rain can damage plants, destroy soil and water resources. We often hear about the death of fish or amphibians in acidic lakes or ponds—as in the mystery scenario in this GEMS guide. The water in these lakes comes from direct rainfall, from runoff water from the nearby soil, and water from the water table. Because water is their total environment, aquatic animals are among the first to show ill effects when the water becomes too acidic. They act as the "canaries in the mine" to warn us of changes in the acid environment of a lake. Different aquatic organisms have different sensitivities. Even when adult organisms seem to survive in an acid environment, their reproductive cycles can be interrupted or impaired, and fish eggs, newly hatched fish, and other new life can be destroyed. Acid rain also mobilizes trace metals. Trace metals are metals found in trace amounts in the soil or water. When more of these metals dissolve in the water, the amounts in the environment are no longer trace amounts. In certain cases, this may reach toxic levels and poison life.

## How are acids neutralized? What's a buffer? What's buffering capacity?
Due to the chemically opposite nature of acids and bases, they can neutralize each other. That's why people with "acid stomachs" or heartburn often drink a solution of baking soda (a weak base), or take a commercial "antacid." (Antacids are also basic.) Buffers are special chemicals that change the pH of a solution to a different, specific, fixed pH. Different buffers are used to stabilize the pH of solutions at different pH values. Buffers are added to fish tanks to help maintain the water at the specific pH best suited to the requirements of the fish.

Soils and rocks often have a natural buffering capacity— the ability of a soil to resist changes in pH. The greater the buffering capacity a soil has, the more acid is necessary to cause a change in that soil's pH. Lakes have their own buffering capacities, different but related to the buffering capacity of the surrounding rocks and soil. A soil's pH and its buffering capacity affect the pH of a lake because a lake's substrate (surrounding soil) is in constant contact with lake water and because the runoff water that feeds into a lake often filters through the substrate. Lakes also have other characteristics that affect their buffering capacity, such as vegetation and animal life. A lake's buffering capacity can be *depleted*. When more acidity shows up in a lake, the natural buffering capacity of the lake's substrate has been used up. When a lake's buffering capacity is exhausted, and its pH drops well below 6, the lake water becomes high in sulfate, aluminum, and other ions, and there is a vastly reduced amount of life in the lake. Buffering capacity can also be *renewed*, by such things as the weathering of minerals in the lake's surroundings.

## How does acid rain affect people?

Although acid rain does not eat through clothing or dissolve skin, it affects people in many ways. It affects water quality, which we use for drinking and washing. It affects the soil, which we need to grow food and to make building materials. Acidity can also dissolve trace metals (mercury, aluminum, cadmium, lead, and asbestos) in the atmosphere, the soil, and in water pipes. It can corrode pipes. It erodes buildings, damaging centuries of human work. It dissolves marble statues, destroying artistic masterpieces. Acid rain often falls far from its polluting source. Once deposited into the atmosphere, airborne gaseous pollutants are often blown hundreds of miles. Depending on wind speed, direction, and duration, pollutants can stay aloft for four days or more. How far the pollutants travel is also affected by how high they are deposited in the atmosphere. The tallest smokestacks can deposit the pollutants high in the atmosphere, where they get carried long distances by the jet stream and other powerful winds.

## What is being done to alleviate the problem of acid rain?

A wide variety of solutions to the problem of acid rain have been attempted. Some address only the symptoms; others try to go to underlying causes. With each solution, costs and benefits must be weighed and funds found. An early (and self-serving) solution sought by industrial polluters was to build taller smokestacks, so pollutants could be carried further away by winds high in the atmosphere. This might have relieved local pressure, but did not address the real issues. Other proposed solutions include: adding buffer to lakes; developing acid-resistant fish(!); increasing use of coal low in sulfur compounds; removing sulfur and nitrogen compounds from fuels and emissions; substituting other energy alternatives for fossil fuels; and conserving energy.

# Sediments

### How do you test for sediment?  What is a Secchi disk?

The word "turbidity" is used to describe the amount of sediment suspended in water.  A Secchi disk is an instrument used to measure turbidity.  An actual Secchi disk is an 8" diameter black and white metal disk with the pattern shown below.  The disk is suspended from a chain or rope that is marked in foot and inch increments.  You can make your own Secchi disks, or order them from a chemical supply company.  The disk is lowered until it disappears, just as the students did in their simulated test.  The feet and inches of depth are recorded.  The disk is then lowered a second time to a depth below where it disappears.  It is then slowly raised until it reappears, and the feet and inches of depth are again recorded.  The two numbers are then added and divided by two.  This average is the turbidity level.

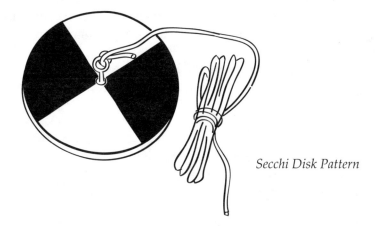

*Secchi Disk Pattern*

A Secchi disk cannot be used if the water is too shallow or if the turbidity of the water is too low.  If used in a fast current, a weight may need to be attached to keep the disk from drifting downstream.  A turbidimeter is another scientific instrument—an optical device that measures the scattering of light, providing a relative measure of turbidity.

A turbidity increase occurs when solids suspended in the water reduce the transmission of light.  These suspended solids can be clay, silt, plankton, industrial wastes, or sewage.  High turbidity may be caused by soil erosion, waste discharge, urban runoff, large numbers of bottom feeders (such as carp) that stir up sediments, or algae.  High turbidity may change the color of water from nearly white to brown or to green from algal blooms.

# Deer Lion Population Simulation

### What is meant by population dynamics?

Populations of organisms are subject to a variety of internal and external factors within the larger ecosystem of which they are a part. Analysis of these factors is part of the study of population dynamics—looking at how and why populations of organisms change. Without limiting factors, populations of animals would keep increasing unchecked. It has been estimated, for example, that one pair of houseflies would have more than five trillion descendants at the end of one year—about 1000 for each person on Earth! Similarly, by the end of a year, one bacterium dividing every 15 minutes would yield a ball of bacteria the size of the known universe expanding outward at the speed of light! This estimated capacity for growth, in the absence of other factors, is called the **reproductive potential** of an organism. But in reality this potential is never unlimited. Such unchecked growth doesn't happen because every organism faces limiting factors. Lack of resources (including food, water, climate, and all the other things an animal needs to survive) is obviously a major factor, as is the impact of predators. In this unit, students explore the impact of varying resources and of predators (mountain lions) on the deer population in Parallel Park.

Populations tend to stabilize at what is called the **carrying capacity,** which represents the balance between reproductive potential and limiting factors. It is quite possible for the carrying capacity of an area to change. If more resources become available (such as when phosphates are added to streams through fertilizers, sewage, and/or laundry detergent) so that algal blooms occur, then the carrying capacity rises, at least temporarily. If a resource is depleted, such as when too many deer eat vegetation and cause erosion, then fewer plants can grow and carrying capacity is lowered, perhaps for years. Given the intricacy of the natural world, the diversity of organisms, and the impact of one of those organisms (humans!) on the environment, the study of the interacting factors that contribute to population dynamics is "naturally" very complex!

# Birdwatching and Water-Related Information

## Birdwatching

### Northern Pintail

The Northern Pintail is the most widely distributed North American duck. They take their name from the male's distinctly pointed tail. Pintails can be found on brackish and freshwater wetlands containing shallow water. They nest in the prairie pothole region (North Dakota and southern Saskatchewan) to the Great Lakes, eastern Canada, and the Arctic and winter primarily in California, Texas, and Louisiana. Pintails eat a variety of grains as well as other seeds, pond vegetation, aquatic insects, crustaceans, and snails. Like Mallards, they are a dabbling duck and feed by tipping tail-up to reach food.

### Northern Harrier

This common (and only) North American harrier is also known as the Marsh Hawk. They are often seen cruising low over open fields or marsh areas with their wings held slightly above the horizontal. Harriers are long-tailed and long-legged. The male is pale gray on the back and whitish beneath while the female is brown on the back and heavily streaked beneath. Both sexes have a very distinct white rump patch at the base of the tail. Harriers hunt by methodically flying low back and forth over fields or marshlands. Their concave facial disk of feathers helps channel sounds to their unusually large ear openings helping them to locate prey. When prey is found, the bird may stall in flight and pounce, or hover for a better look or listen. Harriers eat small mammals (voles, mice, rabbits, moles, rats, and young hares), small birds (song birds, pheasant chicks, and ducklings), as well as frogs, lizards, snakes, fish, and insects.

### Belted Kingfisher

Common throughout almost all of North America, the Belted Kingfisher can often be seen on an exposed perch near a variety of aquatic habitats—rivers, streams, creeks, lakes, ponds, swamps, calm marine waters, and estuaries. They have a stocky body, a ragged crest along the entire head, a strong chiseled bill, and striking blue and white markings. The diet of kingfishers consists primarily of a variety of small fish, as well as mollusks, crustaceans, insects, amphibians, reptiles, young birds, small mammals, and berries. They catch their prey by diving head first into water from their high perch, generally getting most prey near the surface of shallow water. They then come out of the water and fly back to their perch where they stun their prey by smacking it several times against the branch. Then they toss it into the air, catch it head first, and swallow it. Clear water is vital to kingfishers for hunting success. They prefer water that's not overgrown with vegetation.

## Water

### How does water get underground? What is the water table?

When it rains and snows, some of the water stays on the surface of the Earth and some of it seeps underground. This underground water travels through cracks and air spaces until it hits a harder layer of rock underneath which it can't soak through and forms a pool. The top of this underground pool is called the water table. There is a lot more fresh water underground than there is on the surface of the Earth.

### Where does well water come from?

People drill holes down into the ground to the water table (underground water) and pump it up to the surface where it is temporarily stored in a well. This water can be used for drinking and irrigation.

### Why do we treat drinking water?

Most rainwater is clean, but rivers and lakes can be polluted by chemicals, sewage, or bacteria. Water treatment factories take out the chemicals, filter out any solids, and kill harmful bacteria. People add chlorine to drinking water to remove certain kinds of bacteria that can cause sickness. (Not all bacteria are harmful; there are lots of helpful bacteria in our bodies.)

### What is a watershed?

A watershed refers to the land, not to the water. A watershed is all the land that captures water in any form (rain, snow, dew, or hail) and drains into a particular stream system, pond, or lake.

### What are salts? Are salts bad for fish?

Salts are chemicals that dissolve easily in water. The table salt we put on our food is one kind, but there are many others. There are always some salts in any water supply, but too many salts can harm the plants and animals that use the water. Humans need water with some salts in it, but both too many salts dissolved in water (**hard water**) or too few salts (**soft water**) can cause health problems. Some fish live only in salty ocean water and some only live in fresh water, although there are a smaller number that can survive in both. Fresh water fish can die if their water supply becomes too salty.

### How do salts get in the water?
Rocks and soil are made of minerals. Minerals that can dissolve easily in water are called salts. When water flows over the ground and underground it passes over rocks and soils and dissolves the salts from them. Some rocks, such as limestone, dissolve easily, giving up their salts. Calcium, magnesium, and sodium are all salts that get into water.

### What is the difference between fresh and salt water?
Fresh water comes from rain and snow, and is found in rivers, streams, and in the underground water table. Ocean water contains more salts than fresh water, so it is called salt water. If salt water mixes into fresh water, it can kill the fresh water life.

### What is salt water intrusion?
When wells pump too much water up from the water table, the water level gets low. If an ocean is nearby, the salt water from the ocean starts to seep into the water table where the fresh water used to be. This makes the well water salty. This is called salt water intrusion.

# Oil

### Where does oil come from? How does underground oil form?
Some oils, such as cooking oils, come from plants. The oils that can pollute, like car oils, are pumped up from deep underground. Dead plants and animals from many millions of years ago have decayed underground and have been squeezed into a black mucky substance called crude oil. These fuels come from an ancient time when there were dinosaurs and other prehistoric animals and plants, so they are sometimes called "fossil fuels." Crude oil can be refined into motor oil.

### What is oil refining?
Crude oil is taken by truck or ships called tankers to oil refineries, where it is made into different kinds of oil that are ready to use—such as gasoline, kerosene, and motor oil. These oils are called refined oils. Plastics and many other things are also made from oil.

# Sources

## Daphnia

Daphnia should be kept in indirect light at 65–75 °F in an aquarium of aged tap water. You can feed them yeast. Add a pinch of active dry yeast to about 3 oz. of warm water and let it stand for 2–3 hours. Feed a few drops at a time.

If you choose to use live daphnia for the biological test results in Activity 2, you may be able to purchase them at your local aquarium store or order them from the following sources:

**Carolina Science and Math**
2700 York Rd.
Burlington, NC 27215-3398
(800) 334-5551
fax (800) 222-7112
e-mail: carolina@carolina.com
www.carolina.com

    item # BA-14-2314
    daphnia
    $6.20 for enough for 30 students

**Flinn Scientific**
P.O. Box 219
Batavia, IL 60510-0219
(800) 452-1261
fax (630) 879-6962
e-mail: flinn@flinnsci.com
www.flinnsci.com

    item # LM1109
    large daphnia
    $6.75 for enough for 30 students

**Frey Scientific**
100 Paragon Parkway
Mansfield, OH 44903
(888) 222-1332
fax (888) 454-1417
www.beckleycardy.com

    item # S05685    item # S11266
    daphnia          daphnia
    $5.70 for 25    $12.95 for 100

**Nasco Science**
901 Janesville Ave.
Fort Atkinson, WI 53538-0901
**or**
4825 Stoddard Rd.
Modesto, CA 95352-3837
(800) 558-9595
Wisconsin fax (920) 563-8296
California fax (209) 545-1669
e-mail: info@nascofa.com
www.nascofa.com

    item # LM00039(A)M
    large daphnia
    $5.60 for enough for 30 students

**Sargent Welch**
P.O. Box 5229
Buffalo Grove, IL 60089-5229
(800) 727-4368
fax (800) 676-2540
e-mail: sarwel@sargentwelch.com
www.sargentwelch.com

    item # WL50713
    large daphnia
    $6.89 for enough for 25–50 students

## Small cubes and dice

**Cuisenaire**
P.O. Box 5040
White Plains, NY 10602-5040
(877) 411-2761
fax (877) 368-9033
www.cuisenaire.com

**Didax Educational Resources**
395 Main St.
Rowley, MA 01969-3785
(800) 458-0024
fax (800) 350-2345
www.didaxinc.com

**Learning Resources**
380 North Fairway Dr.
Vernon Hills, IL 60061
(888) 489-9388
fax (800) 222-0249
www.learningresources.com

**Math Learning Center**
P.O. Box 3226
Salem, OR 97302
(800) 575-8130
fax (503) 370-7961
www.mlc.pdx.edu

**Nasco Math**
901 Janesville Ave.
Fort Atkinson, WI 53538-0901
**or**
4825 Stoddard Rd.
Modesto, CA 95352-3837
(800) 558-9595
Wisconsin fax (920) 563-8296
California fax (209) 545-1669
e-mail: info@nascofa.com
www.nascofa.com

# Resources

## Related Curriculum Material

There are a wealth of curriculum materials relating to ecology and the environment. To explore the many units available, peruse resources such as the National Science Resources Center publication *Resources for Teaching Elementary School Science*, National Academy Press, Washington, DC, 1996, and *NSTA Pathways to the Science Standards*, Lawrence F. Lowery, editor, National Science Teachers Association, Arlington, Virginia, 1997. Both books include listings and concise annotations of life science, ecology, and environmental units that would nicely complement or extend the activities in *Environmental Detectives*. You will also find many curriculum materials recommended in publications of leading environmental education organizations and a vast array highlighted on the Internet. We welcome your suggestions.

Modules from the **Science Education for Public Understanding Program (SEPUP)** from the Lawrence Hall of Science (LHS), often focus on the interaction between science and society when facing complex environmental issues. For example, in the SEPUP Module, *Investigating Groundwater: The Fruitvale Story*, the aquifer feeding a well is found to be contaminated. A hands-on simulation illustrates how the source and the extent of the contamination can be determined. Students take the roles of concerned professionals and community members and try to develop a clean-up strategy. The year-long SEPUP courses also reflect a wide-ranging environmental concern. The courses are: *Science and Life Issues* (Grades 7–8); *Issues, Evidence and You* (Grades 8–9); and *Science and Sustainability* (Grade 10). SEPUP is available from Lab-Aids, Inc., (800) 381-8003.

The **Full Option Science System (FOSS),** also developed at LHS, includes an *Environments* unit for Grades 5 and 6 with investigations to develop concepts of environmental factor, tolerance, environmental preference, and environmental range. There are also many excellent environmental activities in the bite-size modules of the LHS-originated **Outdoor Biology Instructional Strategies (OBIS)** program. The **Science Curriculum Improvement Study 3 (SCIS3)** includes a *Populations* unit in which students explore the interrelationships of organisms and their populations. FOSS, OBIS, and SCIS3 are available from Delta Education, (800) 258-1302.

The **Science and Technology for Children (STC)** series, from the National Science Resources Center, the Smithsonian Institution, and National Academy of Sciences, has an Ecosystems unit for Grades 5–6, and other related units, available from Carolina Biological Supply Company, (800) 334-5551.

**Adopt-A-Watershed**
P.O. Box 1850
Hayfork, CA 96041
(530) 628-5334
www.adopt-a-watershed.org

Adopt-A-Watershed is a K–12 school/community learning experience which uses a local watershed as a living laboratory. Students engage in hands-on activities, making science applicable and relevant to their lives. It engages students at each grade level in five important elements: applying science concepts directly to a local watershed; monitoring local watersheds through field studies; restoring watersheds through community needs based projects; educating through community action projects; and reflecting upon concepts learned and contributions made to the community.

**Canoes in Sloughs**
The Watershed Education Program of Save The Bay
1600 Broadway, Suite 300
Oakland, CA 94612
(510) 452-9261

Canoes in Sloughs puts San Francisco Bay Area middle and high school students directly on the water in canoes to explore the estuary ecosystem in all its diversity. The field trip brings San Francisco Bay to life for students, helps them to develop self-confidence and teamwork skills, enhances the effectiveness of in-class instruction, and positively affects students' attitudes toward science and ecological concepts. Teachers who schedule a field trip or attend a teacher training trip receive a curriculum guide with activity ideas for the classroom. The curriculum materials are best used as a preparation or follow-up to a field trip, but can also stand alone. The activities in the guide are meant to heighten student awareness of local environmental issues, specifically those that have an impact on water quality.

**Earth Resources—A Case Study: Oil**

This 6th–12th grade integrated science curriculum features over 18 lessons that enable students to learn about the life cycle of a natural resource (using oil as a case study), how it is formed, discovered, extracted, processed, used, collected, and recycled. Through hands-on laboratory investigations using a variety of science concepts from earth and natural sciences, students will identify the environmental impacts of using a natural resource and the positive actions they can take to protect the environment. The curriculum was developed through a partnership between the California Scope, Sequence and Coordination Project (SS&C)/K–12 Alliance, the Integrated Waste Management Board, and the Department of Education guided by an Advisory Board of science educators and experts from oil and oil recycling industries, environmental groups, and state government. The curriculum is nine to twelve weeks long, and is based on the State Science Framework and National Science Education Standards. California teachers can attend one of the workshops being offered statewide. For more information, either call the SCATS Center at (916) 278-4785 or contact Joanne Vorhies at the Integrated Waste Management Board at jvorhies@ciwmb.ca.gov or (916) 255-2362.

**Project WILD**
National Office
707 Conservation Lane
Suite 305
Gaithersburg, MD 20878
(301) 527-890
www.projectwild.org

Project WILD is a conservation and environmental education program. Through a national network of state coordinators and facilitators, Project WILD provides workshops and materials focusing on hands-on, activity-based, environmental education. Their two main activity guides, the *Project WILD K–12 Activity Guide* and the *Aquatic Education Activity Guide* are not for sale; rather they are available to teachers and other educators who have attended workshops offered by coordinators in each state. For more information contact the Project WILD national office or visit their web site.

**Pond and Brook: A Guide to Nature in Freshwater Environments**
by Michael J. Caduto
University Press of New England,
Hanover, New Hampshire. 1990

Contains information and hands-on activities for teachers and students to investigate freshwater environments. Designed for the amateur naturalist, this is a valuable tool for teaching about the unique properties of water, the basic principles vital to understanding aquatic life, and the origin of freshwater habitats.

**WOW! The Wonders of Wetlands: An Educator's Guide**
The Watercourse
P.O. Box 170575
Montana State University
Bozeman, MT 59717

This is a comprehensive guide for developing a wetlands study program appropriate for grades K–12. Each activity features clear headings detailing grade levels, skills, materials, field or lab procedures, and appropriate assessment techniques. The guide also gives details on how to plan and develop your own wetland habitat.

The **Aquatic Outreach Institute** conducts the *Kids in Creeks, Kids in Marshes, Kids in Gardens,* and *Watching Our Watersheds* workshops for educators in Alameda, Contra Costa, and Marin counties (in the San Francisco Bay Area). These workshops offer resources and activities to assist educators in teaching about the importance of protecting local watersheds. Alumni of these workshops can apply for grants to provide supplies and equipment for creek or marsh studies for their students. The Aquatic Outreach Institute also sponsors an annual conference focused on watershed education. For more information contact:
Dede Sabbag, Education Program Coordinator
Aquatic Outreach Institute
155 Richmond Field Station
1327 South 46th Street
Richmond, CA 94804
(510) 231-5784

## Catalogs of Relevant Material

**Acorn Naturalists**
17300 East 17th Street, #J-236
Tustin, CA 92780
(800) 422-8886
www.acornnaturalists.com

This catalog is filled with resources for exploring the natural world. It contains a myriad of materials such as field guides, equipment, puppets, and a large variety of books—including several on understanding and developing environmental education programs.

**Environmental Media Corporation**
P.O. Box 99
Beaufort, SC 29901-0099
(800) 368-3382
www.envmedia.com

This catalog contains many environmental education resources for the classroom and the community. It includes books, videos, CD-ROMs, and equipment. The catalog is conveniently arranged by topics such as biodiversity, forests, microlife, and environmental stewardship.

## Books

*Earth Book for Kids: Activities to Help Heal the Environment,* Linda Schwartz, Learning Works, 1990.

Creative ideas with easy-to-follow instructions show kids how to make their own paper, compare phosphate levels in detergents, test the effects of oil pollution, conduct a recycling survey, create a trash sculpture, redesign a package, chart a flush, measure acidity, and make a difference in many other exciting ways.

*Fifty Simple Things You Can Do to Save the Earth,* Earth Works Group, Earth Works Press, Berkeley, California, 1989.

Offers everyday solutions for combating environmental problems, such as alternatives to household hazardous wastes, stopping delivery of junk mail, use of energy-efficient light bulbs and shower heads, and recycling motor oil.

*Kid Heroes of the Environment*, Earth Works Group, Earth Works Press, Berkeley, California, 1991.

An inspiring collection of environmental success stories about real kids who are doing great things for the earth. Each story contains a description of a kid-based project, interviews, and a listing of sources readers can contact for further information.

*Janice VanCleave's Ecology for Every Kid: Easy Activities That Make Learning Science Fun*, Janice VanCleave, Wiley & Sons, New York, 1996.

*Our Endangered Earth: What We Can Do to Save It*, John Langone, Little Brown, Boston, 1992.

Discusses the environmental crisis, focusing on such problems as overpopulation, the pollution of water, air and land, ozone depletion, global warming, and disappearing wildlife. Suggests ways to improve life in the twenty-first century.

*Rivers, Ponds, and Lakes*, Anita Ganeri, Dillon Press, New York, 1992.

Part of the Ecology Watch series, this book describes how modern life is affecting ponds, rivers, and lakes globally and discusses possible ways of saving the endangered species in these waterways.

*Small Worlds: Communities of Living Things*, Howard E. Smith, Jr., Charles Scribner's Sons, New York, 1987.

Describes a number of small self-contained communities, including a sand dune, tidal pool, old barn, and vacant lot, and examines the ways in which the plants and animals interact.

*The Next Step: 50 More Things You Can Do to Save the Earth*, Earth Works Group, Andrews & McMeel, Kansas City, Missouri, 1990.

# Videos

**The Environment Series: The Mystery of the Dead Fish; The Case of the Mysterious Neighbor; The Mystery of the Statue; The Missing Ingredient; The Lost Book Report Caper; The Power Puzzle**

Produced by the Missouri Botanical Gardens, this collection of videos examines how ecosystems work and the importance of each component of the environment. The components of the global ecosystem, air and air pollution, water and water pollution, biological resources, energy sources and conservation, and solid waste are featured. Each 20-minute video presents the same four students with a different mystery to solve. The videos are accompanied by a teacher's guide suggesting how to use the tape effectively. The guides also present the clues in detail, the science concepts explored, discussion topics and topic summaries, and follow-up hands-on activities. For more information, call MBG Videos (314) 531-1100 or (800) 927-9229.

## Plankton Play

As part of a series on microlife for middle school students, this video shows copepods, daphnia, and volvox. The video features a crew of micronauts in a very very tiny submersible exploring an open-water environment. When their craft collides with a daphnia, they get a unique view of the animal's internal organs in action and its special adaptations for living in suspension. The video summarizes the adaptations of these organisms for open-water life, their reproduction , and the roles they play in the pond ecosystem. Available from BioMedia Associates (888) 248-6665 or Carolina Biological Supply Company (800) 334-5551.

# Interactive CD-ROM

## Investigating Lake Iluka: A Simulated Lake Environment Designed to Support the Teaching of Ecology

Published by Interactive Multimedia Pty Ltd. in 1993 and designed for grades 6–10, this multimedia CD for Macintosh or Windows simulates a fictitious coastal lake. The software contains tools for investigating the physical, chemical, and biological characteristics of the ecosystem—including many of the same tests as in *Environmental Detectives*. It also includes a variety of social issues that are typical of such an environment and a user notebook for collecting information. Case studies of ecological scenarios are posed directly to the user through media reports of problems occurring in the lake. Each scenario can be investigated using the common scientific tools. Users develop a broad array of scientific investigation skills using this realistic simulation. The set of built-in problems range from simply reporting on specific information, making measurements, or comparing ecosystems to proposing a solution to a problem which is threatening the well being of the lake environment. For more information, contact: The Learning Team, 84 Business Park Drive, Armonk, New York 10504, (914) 273-2226 or call (800) 793-TEAM.

# Watershed Model

Monterey Bay National Marine Sanctuary
299 Foam Street
Monterey, CA 93940
(831) 647-4201
www.mbnms.nos.noaa.gov/

The Monterey Bay National Marine Sanctuary has a hands-on urban runoff educational model available for use by schools and other organizations. The plastic model imitates a typical watershed—with residential areas, a golf course, a farm, a logged forest, a factory, and a sewage treatment plant plus an area in which water accumulates. Students place various pollutants on the model in appropriate places using things like cocoa for sediment, koolaid of various colors for pesticides, fertilizer, and industrial waste, and chocolate sprinkles for animal waste. Then students apply rain from spray bottles and experience the fouled water entering the previously clear sanctuary water. For information on how to check out the model from the Sanctuary office, contact Maris Sidenstecker at (831) 647-4216. To see a picture of the model, go to www.mbnms.nos.noaa.gov/Resourcepro/watermodel.html

# Assessment Suggestions

## Selected Student Outcomes

1. Students increase their content knowledge about the causes and effects of water pollution—including chlorine, erosion/sediment, acid rain, predator/prey relationships, phosphate pollution/algal blooms, and oil pollution.

2. Students improve in their ability to conduct research and develop other inquiry and literacy skills through conducting simulated tests and investigating files of written/ pictorial materials and data. They are able to summarize their findings and determine which results and files are most relevant to the problem at hand.

3. Students are able to distinguish between evidence and inference, to defend conclusions by citing evidence, and to refrain from jumping to conclusions until sufficient data is analyzed.

4. Students deepen their understanding of the complex interactive nature of many environmental problems and recognize that there are many "gray areas." They are able to discuss a range of possible solutions and, through group debate, compromise, and problem solving, select their own solutions and explain their reasoning.

## Built-In Assessment Activities

**Letter-to-the-Editor Special Assessment** (see page 210). In this assignment, students read a "letter to the editor" blaming Don Juan Tunó for the fish die-off. The letter has inaccuracies for students to identify as they write a letter of rebuttal. Student responses can help assess their understanding of the unit's science content, their ability to distinguish evidence from inference, their research and communication skills, and their awareness of the complexity of environmental problems. Helpful information for the teacher related to this assessment is provided on pages 209 and 211. (Outcomes 1, 2, 3, 4)

**Main Trunk of Activities.** In most class sessions, students perform simulated tests or study data gathered by others, do research in files, discuss their findings in groups and with the whole class, then reflect on the possible cause of the fish dying, adjusting previous predictions as needed. The teacher can observe both individual student and small group participation to gauge student progress and growth in applying content knowledge, reading for comprehension, doing effective research, weighing evidence, responding to diverse ideas, engaging in group problem solving, and explaining their reasoning. (Outcomes 1, 2, 3, 4)

**Detective Notebooks.** Each activity concludes with an opportunity for students to reflect and to write in their Environmental Detective notebooks. The notebooks can be used by the teacher for assessment purposes. In addition to reflecting growth in student knowledge and skills over the course of the unit, the notebooks can also reveal the level of student understanding of the complexity of the mystery, their ability to express and explain their ideas in writing, and how well they can adapt and modify previous predictions in light of more evidence. (Outcomes 1, 2, 3, 4)

**Focus on Research Skills.** The research students conduct in the "files" provides many opportunities to deepen understanding, improve reading comprehension, and make numerous connections to the mystery. Students gain practice perusing numerous documents, deciding which are pertinent and why, and then gleaning information they need. In small group and class discussions, the teacher can observe how well students have assimilated and applied the research information. Most students need encouragement, practice, and guidance to develop and refine their research skills. You may want to give special emphasis to this aspect of the unit by frequently making students conscious of what they are doing and developing some guidelines for helping students evaluate their own research methods. We'd welcome hearing your ideas and any comments on improvement in your students' research abilities. (Outcome 2)

**Conclusions and Solutions.** In Activity 8, students change from the role of environmental detectives into that of a local decision-making board. They brainstorm solutions to the problems they think most likely to be killing the fish, debate each solution, and vote on it. The solutions of each group, and their detailed explanations regarding the factor(s) they think most responsible and how problems might be solved or ameliorated can help assess overall learning throughout the unit. (Outcome 4)

## Additional Assessment Ideas

**Optional Homework.** For many sessions, optional homework question sheets are included, to encourage students to process information on their own. These sheets can help you learn more about specific content knowledge students have gained, as well as provide insight into what is going on in the discussion groups. If you assign all the home-work sheets, you will have a fairly complete record of student responses concerning the specific scientific and environmental information in the unit. (Outcome 1)

**Local Environmental Issues.** Have individual students or small groups examine a current environmental situation or issue in the local region—preferably somewhat controversial. Student groups could locate and read several articles, find information on the web, do a few phone interviews, invite a representative of an environmental organization to class, find other sources and summarize what they've found out in a written report and a classroom presentation. The presentation could be a role-play of a report to the City Council, to convince the council to take action. (Outcomes 2, 4)

**Make-A-Mini-Mystery.** Encourage students to devise their own ideas for an environmental mystery scenario. This can be in outline form, and should not be as extensive or complex as the mystery in the unit. The scenario should focus on a major environmental and/or health-related problem connected to pollution, toxic waste, erosion, pesticides, radiation, or other issues. Students could exchange scenarios and discuss "suspects." The outlines could be reviewed by the teacher on various criteria, such as originality/creativity, how well the environmental problem is described, and/or how well the scenario indicates student understanding of interacting factors and "gray areas." (Outcomes 1, 4)

# Literature Connections

**The Big Book for Our Planet**
edited by Ann Durell, Jean Craighead George, and Katherine Paterson
Dutton Children's Books, New York. 1993
Grades: 4–8

Nearly thirty stories, poems, and non-fiction pieces by notable authors and illustrators, demonstrate some of the environmental problems now plaguing our planet such as overpopulation, tampering with nature, litter, pollution, and waste disposal.

**Catch of the Day: The Case of the Helpless Humpbacks**
by Emily Lloyd
McGraw-Hill, New York. 1997
Grades: 3–7

This is part of the Kinetic City Super Crew series, based on the public radio show produced by AAAS. The Super Crew is a group of 7 young people who travel the world solving problems. In this book, they must try to prevent whales from getting caught in fishing nets and drowning off the coast of Nova Scotia. In the process, the readers learn much about whales and their plight. Visit the Super Crew at their web site http://www.kineticcity.com/

**The Eleventh Hour**
by Graeme Base
Harry N. Abrams, New York. 1988
Grades: 3–8

This uniquely illustrated picture book is about an elephant's eleventh birthday party with other animals as guests. In addition to being an illustrative and poetic *tour de force,* this book is a compelling mystery. We learn that one of the animals has managed to gobble up the special birthday banquet. All eleven animals are suspects, and the solution is said to be contained in the many layers of clues provided throughout the book. The end of the book is sealed so you can't find out who did it until you think you have it solved!

### Forest Slump: The Case of the Pilfered Pine Needles
by Emily Lloyd
McGraw-Hill, New York. 1997
Grades: 3–7

This is part of the Kinetic City Super Crew series, based on the public radio show produced by AAAS. The Super Crew is a group of 7 young people who travel the world solving problems. In this first book, they receive a call reporting pine needles missing from a forest in South Carolina. As they investigate the scene of the crime, they learn quite a bit about the forest ecosystem. A good blend of humor and learning. Visit the Super Crew at their web site http://www.kineticcity.com/

### From the Mixed-Up Files of Mrs. Basil E. Frankweiler
by E. L. Konigsburg
Dell, New York. 1977
Grades: 5–8

Twelve-year-old Claudia and her younger brother run away from home to live in the Metropolitan Museum of Art and stumble upon a mystery involving a statue attributed to Michelangelo.

### Incognito Mosquito Makes History
by E. A. Hass; illustrated by Don Madden
Random House, New York. 1987
Grades: 4–7

The famous insective travels back in time to solve five mysteries involving such notables as Christopher Columbug, Benetick Arnold, Buffalo Bill Cootie, Tutankhamant, and Robin Hoodlum.

### Incognito Mosquito, Private Insective
by E. A. Hass
Lothrop, Lee & Shepard Books, New York. 1982
Grades: 4–7

In this book, the first of several, the mosquito detective tells a cub reporter of his exploits and encounters with such insect notables as Mickey Mantis, F. Flea Bailey, and the Warden of Sting Sting Prison. In each chapter, the detective tells of a past case whose solution is at first left for the reader to solve; the final page of the chapter then gives the solution.

## The Missing 'Gator of Gumbo Limbo: An Ecological Mystery
by Jean C. George
HarperCollins, New York. 1992
Grades: 4–7

Sixth-grader Liza K and her mother live in a tent in the Florida
Everglades. She becomes a nature detective while searching for
Dajun, a giant alligator who plays a part in a waterhole's oxygen-
algae cycle, yet is marked for extinction by local officials. She
studies the delicate ecological balance in order to keep her outdoor environment beautiful. This "ecological mystery" combines
precise scientific information and a variety of important environmental concerns with excellent characterization, a strong female
role model, and an exciting, complex plot.

## Motel of the Mysteries
by David Macaulay
Houghton Mifflin, Boston. 1979
Grades: 6–Adult

Presupposing that all knowledge of our present culture has been
lost, an amateur archaeologist of the future discovers clues to the
lost civilization of "Usa" from a supposed tomb, Room #26 at the
Motel of the Mysteries, which is protected by a sacred seal (a "Do
Not Disturb" sign). This book is an elaborate and logically
constructed train of inferences based on partial evidence, within a
pseudo-archaeological context. Reading this book, whose conclusions they know to be askew, can encourage students to maintain
a healthy and irreverent skepticism about their own and other's
inferences and conclusions, while providing insight into the
intricacies and pitfalls of the reasoning involved. This book can
help deepen the practical experiences students have gained in
distinguishing evidence from inference. It also helps demonstrate, in a humorous and effective way, the connection between
detective work and the science of archaeology.

## The River
by David Bellamy; illustrated by Jill Dow
Clarkson N. Potter/Crown, New York. 1988
Grades: 3–5

Plants and animals coexist in a river and have to struggle for
survival when a man-made catastrophe strikes. Details about
stream ecology include a description of the effects of waste water
discharged from a factory and how the bacteria, algae, and
oxygen interact in the dam area and beyond. The ending seems
overly optimistic with the river "back to normal" a month after
the waste was discharged. "Everyone hopes the factory owners
will be more careful in the future."

### Who Framed Art Decco?
by Margaret Benoit
McGraw-Hill, New York. 1997
Grades: 3–7

On a rainy Saturday, homicide detective Angel Cordoni decides to visit a local art gallery to view works of sculptor Bern T. Sienna and painter Forrest Greene. But her day off becomes a working day when she learns someone is stealing artwork from the gallery and trying to frame gallery owner Arthur Decco for insurance fraud. In addition, art critic Rave N. Phoole is killed and it appears a piece of one of the sculptures is the murder weapon. Through brilliant deduction, careful experimentation, and applied science, the clever detective solves the case. The book is a well-blended combination of entertainment and learning.

### Who Killed Olive Souffle?
by Margaret Benoit
McGraw-Hill, New York. 1997
Grades: 3–7

At a snow-bound country inn, homicide detective Angel Cardoni works to solve the murder of the inn's famous chef using the only forensic tools available—the inn's kitchen supplies. Despite the lack of evidence and a number of red herrings, Angel uses many important science-based crime-investigation techniques (such as paper chromatography and identifying liquids by their boiling point and density) to solve the case.

### Who Really Killed Cock Robin?
by Jean C. George
HarperCollins, New York. 1991
Grades: 3–7

A young hero in this compelling ecological mystery examines the importance of keeping nature in balance. This is an inspiring account of an environmental hero who becomes a scientific detective. With its interweaving factors that contributed to Cock Robin's death, this book was an inspiration for the development of *Environmental Detectives*.

# Summary Outlines

## Activity 1: Introducing the Mystery

### Getting Ready

#### Before the Day of the Activity

**Setting Up Headquarters**
1. Prepare Suspect Chart.
2. Make copy of pictures/statements for Juan Tunó, Don Juan Tunó, and Avery Wun.
3. Make overhead of Timeline and make large map.

**Preparing for Teamwork**
1. Group students.
2. Prepare clue cards.
3. Plan for notebooks and copy pages for them.
4. Read over Clue Cards, map, and statements.

#### On the Day of the Activity
1. Tape map and Suspect Chart to a wall.
2. Have rest of materials on hand.

### Introducing the Mystery of the Dying Fish
1. Tell students they will be learning about environmental science.
2. Show large map.  It is not a real area, but is based on actual places.
3. Explain problem.  A fish die-off has been going on for five years.  Fish autopsies are pending, but results won't be back for a month.
4. Students will try to solve this **environmental mystery.**  Like real scientists, they'll conduct tests and research.
5. Ask what a detective does.  Students will be scientists and "environmental detectives."
6. Lead discussion about needs of fish; factors that impact these needs; how a fish kill might affect other organisms and the environment; how students could find out what's going on; what might be killing the fish.

### Creating a Timeline of the History of the Gray Area
1. Tell students they will first look for clues in **area's history.**
2. Distribute map, timeline, and lined paper for notebooks.
3. On overhead, model using clue cards to fill in timeline.
4. Explain that last 20 years of top timeline are expanded on bottom.
5. Each student records events on timelines, but group shares clue cards.  Have a student from each group get cards and have groups begin.
6. Circulate, encouraging discussion.  When students finish, have them return cards and reassemble.

## Timeline Discussion

1. Ask students if they found clues that might explain why fish are dying. Accept all ideas, say Board of Supervisors suspects water slide and plans to close it down. In response, a student, Juan Tunó, has started his own investigation.

2. Show Juan's picture, read his statement, then tape it next to Suspect Chart. Emphasize that he is **not** a suspect.

3. Say that Juan's uncle, Don Juan Tunó, has a different opinion. Show his picture, ask a student to read his statement, then tape it on Suspect Chart. Ask students for comments.

4. Say that Juan Tunó has persuaded Board not to shut down water slide yet, but Board needs to decide in _____ *(however long unit will last)*.

## Outlining the Unit for the Students

1. Ask students what they think could be killing the fish. Accept all ideas and list. Make sure list includes chlorine, acid rain, dirt (or sediment), phosphates, and oil. Put a star next to these five substances.

2. Say that where chart and map are located is called "Headquarters."

3. Ask students to be patient; it may take time to make a final decision.

4. There is another suspect. Show picture of Avery Wun, ask a student to read statement, then tape on chart.

5. Give students time to write in notebooks and list questions they have that could help solve the mystery.

# Activity 2: Chlorine Tests

## Getting Ready

### Before the Day of the Activity

1. Remind students to bring notebooks to every lesson.

2. Make four copies of record of Meeting of the Board of Supervisors, and a copy of Water Flea Diagram.

3. Make one set of Chlorine Files per group and put in labeled envelope.

4. Copy and cut Chlorine Discussion Cards—one card per group.

5. Copy Juan Tunó's Chlorine statement and tape under his picture and first statement.

6. Copy Ken Unballe's picture and Chlorine statement and tape on chart.

7. If using, make copy of Chlorine: What Do You Think? sheet for each student.

### Introducing the Chlorine Problem

1. Announce that the Board met recently about the water slide's role. You have transcripts. Ask four students to read the parts.

# Get Connected – Free!

### Get the *GEMS Network News*,

### our free educational newsletter filled with...

- **updates** on GEMS activities and publications
- **suggestions** from GEMS enthusiasts around the country
- **strategies** to help you and your students succeed
- **information** about workshops and leadership training
- **announcements** of new publications and resources

*Be part of a growing national network of people who are committed to activity-based math and science education. Stay connected with the* **GEMS Network News.** *If you don't already receive the* **Network News,** *simply return the attached postage-paid card.*

*For more information about GEMS call (510) 642-7771, or write to us at GEMS, Lawrence Hall of Science, University of California, Berkeley, CA 94720-5200, or gems@uclink4.berkeley.edu.*

*Please visit our web site at www.lhsgems.org.*

GEMS activities are effective and easy to use. They engage students in cooperative, hands-on, minds-on math and science explorations, while introducing key principles and concepts.

Great Explorations in Math and Science
**LHS GEMS**

More than 70 GEMS Teacher's Guides and Handbooks have been developed at the Lawrence Hall of Science — the public science center at the University of California at Berkeley — and tested in thousands of classrooms nationwide. There are many more to come — along with local GEMS Workshops and GEMS Centers and Network Sites springing up across the nation to provide support, training, and resources for you and your colleagues!

www.lhsgems.org

## Yes!

Sign me up for a free subscription to the

# GEMS Network News

filled with ideas, information, and strategies that lead to
Great Explorations in Math and Science!

Name_____

Address_____

City_____ State_____ Zip_____

How did you find out about GEMS? (Check all that apply.)
❑ word of mouth ❑ conference ❑ ad ❑ workshop ❑ other: _____
❑ In addition to the *GEMS Network News*, please send me a free catalog of GEMS materials.

GEMS
Lawrence Hall of Science
University of California
Berkeley, CA 94720-5200
(510) 642-7771

# Ideas◄
## *Suggestions*◄
## Resources◄

that lead to Great Explorations
in Math and Science!

**LHS GEMS**

101 LAWRENCE HALL OF SCIENCE # 5200

1-61571-25775-62-X

NO POSTAGE
NECESSARY
IF MAILED
IN THE
UNITED STATES

## BUSINESS REPLY MAIL
FIRST-CLASS MAIL    PERMIT NO 7    BERKELEY  CA

POSTAGE WILL BE PAID BY ADDRESSEE

UNIVERSITY OF CALIFORNIA BERKELEY
GEMS
LAWRENCE HALL OF SCIENCE
PO BOX 16000
BERKELEY  CA  94701-9700

*Get Connected!*

www.lhs...

2. Provide student playing Juan Tunó with Water Flea Diagram and an envelope with the Chlorine Files. Have students come to front and begin.
3. Point out that statements for Ken Unballe and Juan Tunó are posted.

## Small Group Study and Discussion
1. Tell students they'll get copies of documents the Board received.
2. In Chlorine File they'll find: results of tests; Juan Tunó's clippings; e-mails, notes, and documents from envelope left near his door.
3. Groups read documents and search for clues. When finished reading and discussing, one person gets Chlorine Discussion Card.
4. Distribute Chlorine Files and have students begin.

## Large Group Discussion
1. Refocus the class, and ask one group to summarize results of Juan Tunó's chlorine **biological** test. Ask, "What does the lack of water fleas downstream from the water slide tell us?" "What were the results from the **chemical** test for chlorine?" "Why might the results of the two chlorine tests be different?"
2. Tell students environmental scientists often use **both** biological and chemical tests. If they don't mention, ask them to look at water slide schedule to see when sample was collected for chemical test.
3. Students may think they've already solved the mystery. Point out that it may be complex; encourage them not to jump to conclusions.

## Reflecting on Chlorine
1. Ask students to write about whether chlorine is killing the fish and make notes about suspects in notebooks and under suspect pictures.
2. Have groups reassemble Chlorine Files. File will be at Headquarters for later reference. Collect the files.
3. If you've decided to use, assign Chlorine: What Do You Think? sheet as homework or work in class. Have students put sheet in notebooks.

# Activity 3: Acid Rain

## Getting Ready

### Before the Day of the Activity
1. Prepare pH paper for each group.
2. Use masking tape and the permanent marker to label cups.
3. Prepare test solutions as described in guide.

4. Prepare two soil run-off models, making funnel and container from two-liter bottles. Label one model "Fo and Missterssippi River Area Soil Run-off," and other "Rafta River Area Soil Run-off."

5. Prepare soils for run-off models as described in guide.

6. Copy, color, and, if possible, laminate pH scales.

7. Prepare the Acid Rain Files and Clue Cards, Acid Rain Questions, and copy Test Results sheets.

8. Make one copy of Newspaper Interview for each student.

9. If using, make one copy of Acid Rain: What Do You Think? sheet for each student.

### On the Day of the Activity

1. Keep "Gray Area Rain" cup handy for first session introduction. Copy and have on hand picture and Acid Rain statement for LaToya Faktorie, and Juan Tunó's and Don Juan Tunó's Acid Rain statements.

2. Set up ten testing stations as described in guide.

## Session 1: Introducing Acid Rain

### Reflecting on the Evidence About Chlorine

1. For review, ask students if they think chlorine from Ken Unballe's water slide is killing the fish and why.

2. Ask, "What do we know for sure from studying the Chlorine Files?" (Fewer daphnia downstream than upstream, daphnia sensitive to chlorine, and Ken Unballe wrote e-mails). This is **factual** evidence.

3. Help students distinguish between factual evidence and their conclusions or *inferences*. It's important to make inferences, but it's also important to be ready to change your mind if new evidence arises.

### Introducing the Acid Rain Problem in the Gray Area

1. Tell class Juan Tunó is still worried about chlorine, but also acid rain. Read his statement then tape it below his picture.

2. Hold up Gray Area Rain cup—all samples Juan took are same **acidity.**

3. Hold up pH paper. It tests if a substance is acidic or not. Dip it in "rain," and compare color of strip to chart. Color scale called pH scale; numbers below 7 acid—the lower the number, the stronger the acid. Say 7 is neutral, 0 to 6 acidic, and 8 to 14 basic.

### Introducing the Tests for Acidity in Gray Area Waterways

1. Ask students if they think acid rain may be making rivers and lakes too acidic for fish. They can find out at testing stations.

2. Point out stations and procedure sheets. Say acid rain seeps from soils into rivers/lakes. Soils are from Rafta and Fo/Missterssippi River areas.

3. Pour half of cup of rain onto dirt in each filter. Some of water that seeps through will be in the bottom of the bottle to test.
4. Students will need to tear pH paper into 10 little pieces.
5. Show Test Results sheet to add to notebooks.
6. They can go to stations in any order. When finished, they can study the map, other information, and/or write in their notebooks.
7. Distribute pH paper and data sheets, and have students begin.
8. When finished, distribute interview from *Synchrony City Chronicle* to read now or for homework. Dismantle or clean up stations.

## Session 2: Research and Discussion

### Discussing Acid Rain and the Results of pH Tests
1. Ask questions about *Chronicle* interview: The pH of normal rain? Two main kinds of acid in acid rain? What causes acid rain?
2. Ask pH level of Gray Area Rain. Ask, "Is that normal or acid rain?"
3. Tell students most fish can't survive in water below pH 5. Have them look in notebooks for test results. Ask, "Do any of the rivers or lakes you tested have a pH level too acidic for fish to survive?"
4. Ask, "Were there any water samples **not** too acidic for fish?"
5. Have students share ideas and questions about test results.

### Small Group Research and Discussion
1. Post Acid Rain Questions.
2. Each group will get the Acid Rain File, with documents and clue cards.
3. Groups can divide up some of the reading. Suggest that each student in a group of four could be responsible for one question.
4. Suggest they all *skim* documents to "inventory" them, then classify by questions—but some documents may not fit into easy categories.
5. Some documents may be interesting but not helpful for the questions. That's how research is. They can make an "extra" pile.
6. Each student should take notes as they do research.
7. When group is done, each student will share findings.
8. Have students get Acid Rain File and begin their research.

### Large Group Discussion
1. When groups finish, use Acid Rain Questions for class discussion.
2. Ask students if they think acid rain or chlorine is killing water fleas near the slide, and if they found out anything else interesting in files.

3. Show picture of LaToya Faktorie, have student read statement, then tape on chart. Ask for comments.

4. Ask student to read statement by Don Juan Tunó, then tape below picture and his previous statement. Ask for comments.

### Reflect, Predict, and Vote

1. Tell students they'll each receive two post-its, write their names on them, and vote on suspects they think "guilty" by placing on chart.

2. There is also a column on chart with question mark for undecided.

3. They can change votes, so shouldn't worry about the "right" answer. Vote is mostly to see how ideas of class change over time.

4. Have students write in notebooks, write "acid rain" on map at what they think is its source, and put post-its under suspects on chart.

5. If using it, assign as homework or class work the Acid Rain: What Do You Think? sheet and have students place in notebooks.

## Activity 4: Sedimental Journey

### Getting Ready

#### Before the Day of the Activity

1. Prepare eight milk cartons for stations. Label cartons as described in guide and completely unfold top of each carton.

2. Make eight Secchi disks as described in guide.

3. Make one copy each of the picture and Sediment statements for Anton Alogue and Elmo Skeeto, and Juan Tunó's Sediment statement.

4. Make eight copies of the Sediment Test Procedure Sheet.

5. Make one set of Sediment Files for each group; put in labeled envelopes.

6. Make copy of Sediments: What Do You Think? sheet for each student.

#### On the Day of the Activity

##### Preparing Materials for the Erosion Demonstration

1. Fill pan with about 1" of water, and set on overhead projector.

2. Cut a triangle in rim of the two plastic cups.

3. Turn cups upside down on paper towel. Scoop up a tablespoon of soil and place it in a mound on one cup.

4. Scoop up another tablespoon of soil. Put on paper towel and mix with grass clippings. Repack tablespoon, place mixture on top of other cup.

5. Have near projector: prepared cups, tablespoon, extra water, suspect pictures/statements for Alogue and Skeeto, and Juan Tunó's statement.

6. Practice erosion demonstration ahead of time.

**Setting Up the Sediment Test Learning Stations**
1. Fill each milk carton about 8" with water.
2. Vary murkiness with chocolate milk as specified in guide.
3. Choose widely spaced locations for stations.
4. Put one milk carton at each station. Place Secchi disk on paper towel next to milk carton. Put Sediment Test Procedure sheet at each station.

## Introducing the Sediment Problem: Demonstrating Erosion with a Model
1. Turn on overhead projector with pan of water on it. Set cups into water; the water represents lake (or river); cups with soil represent hills.
2. Tell students hills have same amount of soil—one is plain, other has grass mixed in.
3. Show students one tablespoon of water. Ask for predictions about when this much water "rains" on each of the hills.
4. Dribble tablespoon of water on each hill. Ask for observations. Explain that water that soaks into and travels through ground is called *ground water.*
5. Ask for predictions about what will happen when you "rain" another tablespoon on each hill. Dribble the water, ask for observations. Continue as needed.
6. Explain that, when it rains a lot, ground gets saturated, so water runs off surface, carrying dirt with it. This is **erosion.** The dirt deposited in the water is called **sediment.**
7. If students notice hill with grass clippings has less sediment in water around it, ask why. Plant leaves and roots can help prevent erosion.

## Juan Tunó and Sedimentation in the Gray Area
1. Bring focus back to mystery. Juan Tunó has noticed dirt in the water. Read his statement, then tape under picture and previous statements.
2. Show Anton Alogue, ask volunteer to read statement, tape on chart.
3. Remind class that logging operations are owned by Tunó Enterprises.
4. Ask students how they think too much dirt in the water might harm fish and other water life.

## Introducing the Sediment Stations
1. Point out that some erosion of sediments into streams and lakes is always taking place in nature, but too much erosion can cause problems. Students will test to see if Gray Area waters have too much sediment.
2. Show students where stations are located. Go over procedure sheets. They record results on data sheet then move to another station.

3. Students should remove disks after each test—place on paper towel.
4. When pairs complete tests, they can reconsider statements and map and/or add to notebooks.
5. Have students begin.

### Small Group Discussion
1. Each group will send one person to get Sediment File. There aren't too many documents, so everyone should read entire file.
2. Each student will also get Sediments: What Do You Think? sheet. They should discuss all questions, then each fill out own sheet.
3. Students get Sediment File. Groups begin as you distribute Sediments: What Do You Think? sheets. Circulate, help any groups having difficulty.

### Large Group Discussion
1. Use questions on Sediments: What Do You Think? sheet for discussion.
2. Ask, "Do any of the bodies of water you tested have too much sediment?"
3. Ask where students think dirt might be coming from and its cause.
4. Ask, "How can too much sediment harm water life?"
5. Introduce hunters and fishers. Show Elmo Skeeto, have student read his statement, tape to chart.
6. Ask students if they have comments. In next session, they'll find out more about Parallel Park.

### Reflect, Predict, and Vote
1. Have students write in notebooks, write "sediments" on their map where they think source is, and adjust post-its on chart as needed.
2. Have students put completed Sediments: What Do You Think? sheets in notebooks.

## Activity 5: Deer Lion

### Getting Ready (for the Active Version)
1. Make sure you understand rules.
2. Choose space where whole class can form two parallel lines.
3. Make large version of Population Game Graph.
4. Decide where you'll display large graph.

### Getting Ready (for the Dice Version)

#### Before the Day of the Activity
1. Read over the explanation of the dice game.
2. Make one copy of Population Game Graph and Random Attack Board per pair of students.

3. Make large version of Population Game Graph.
4. Make overhead of Random Attack Board.

## On the Day of the Activity
1. If using sponge dice, plump them up by lightly spraying with water.
2. Have student sheets, Juan Tunó's statement, overheads, and bags of dice on hand.

## Getting Ready (for the remainder of the activity)
1. Prepare one set of Deer Lion Files per group; put in labeled envelopes.
2. Copy and cut Discussion Cards so there is one card per group.
3. Make one copy of Juan Tunó's Deer Lion statement.
4. Make overhead of Deer and Mountain Lion Populations in Parallel Park.
5. If using, make copy of Deer Lion: What Do You Think? sheet for each student.

## Focusing on Parallel Park
1. Read Juan Tunó's Deer Lion statement; post below prior statements.
2. Ask students what deer (and other organisms) need to survive. Discuss. Define *resources* as the things animals need to survive.
3. Define *population* as a group of one kind of animal. Ask, "How can a population grow?" "How can populations decrease?"
4. Ask, "How might a change in a deer population affect erosion?"
5. Tell class they're going to play a game about changes in deer populations and resources in Parallel Park.

# The Active Version of the Population Game

## Demonstrating How to Play
1. Teach hand signals that represent resources.
2. Ask four students to help demonstrate rules of the game.
3. Have students facing in two lines—one deer; the other resources.
4. Show how to make dot on game graph for number of deer in first year.
5. Have two lines face away from each other, then have them model how game is played as described in guide.
6. Choose student to graph class results during game. The idea is to learn how populations change in nature, so "cheating" will not help. Emphasize they must have hand signal in place *before* turning around.

## Playing the Population Game—The Active Version

1. When students are clear on procedure, move to large space or outdoors. Display graph, have students form lines, and begin game.
2. Periodically ask students to predict what will happen to the deer population in next round by voting thumbs up (increase), down (decrease), or horizontal (remain about the same).
3. Continue playing for about 10 or more rounds until the "boom and crash" pattern is evident on graph.

## Introducing How to Play the Game with a Predator— The Active Version

1. Tell students a *predator*—a mountain lion—will be introduced to see impact on deer population. You will secretly appoint one person in resource line to be a mountain lion.
2. Graphing volunteer records with different color dot for mountain lions.
3. As before, any deer not matched with resources die and become resources in next round, and resources captured by deer become deer in next round. If deer grabs a mountain lion, lion will roar and grab deer back. Once "attacked" by lion, deer dies and joins resource line.
4. You will choose different people each round to be lions—only for that round. For every lion that catches a deer, two are hidden in resource line next round.

## Playing the Game with a Predator and Graphing— The Active Version

1. Have the students form lines, with number of deer and resources they had in the last round. Students make hand signal while you walk behind resource line and tap a student to make her into mountain lion.
2. Begin game. After each round, volunteer records number of deer and mountain lions. Ask lions who *did* catch a deer in that round to raise hands. Double that number to make lion population for next round.
3. Before each round, tap appropriate number for new lion population.
4. If a lion isn't picked by a deer, the lion gets no meal, dies, and becomes a resource. If no lions are left, start next round with one lion.
5. Periodically ask class to predict whether population of deer or mountain lions will go up, down, or stay same.
6. Once new pattern established on graph, you can end game.

# The Dice Version of the Population Game

### Demonstrating How to Play
1. Hold up one of dice, and explain what colors (or letters) represent.
2. Choose two volunteers to help model game as described in guide.

### Playing the Population Game—The Dice Version
1. Distribute dice and graphs to student pairs and have them begin. Circulate, helping as needed.
2. Let students continue until graphs of most groups show pattern of sharp increases and decreases in deer population (about 10 rounds).
3. If ending session after discussion of first part of game, collect dice.

### Discussing the Population Game—The Dice Version
1. Have students focus on their graphs.
2. Ask students what happened to deer population in game. A sharp population increase is called a *boom*; a major decrease called a *crash.*
3. Discuss, helping students interpret graphs **for themselves.**

### Introducing How to Play the Game with a Predator— The Dice Version
1. Using class graph, draw a graph similar to students' graphs. Last round should show 10 deer.
2. Tell students they'll keep playing the game, but with a *predator*—a mountain lion—to see what impact this has on deer population.
3. Ask volunteer to help model game, then play one more sample round.

### Playing the Game with a Predator and Graphing— The Dice Version
1. Remind partners they should start with however many deer and resources they had in the last round they played.
2. Use class graph to show how to record mountain lions.
3. Point out dot on graph for 10 deer in last round. Use different colored marker for mountain lion population that year. (One mountain lion.)
4. Show how to record deer and lions in second demonstration round.
5. Distribute dice, Attack Boards, scratch paper, pens, and have students begin. Circulate. Encourage partners to switch roles.
6. Once new pattern is evident on graphs, stop playing and collect dice.

### Interpreting the Graphs from *Either Version* of the Game
1. Review what happened when there were no mountain lions.
2. Ask, "What happened to deer population when there were mountain lions?" and lead a discussion.

### Small Group Discussion
1. Ask each group to send one person to get a Deer Lion File and a Deer Lion Discussion Card.
2. Have groups look over files and discuss questions.

### Large Group Discussion
1. Put graph on overhead. Ask students to interpret it.
2. If needed, say hunting mountain lions started six years ago, and that water in river has harmful levels of sediments.
3. Ask, "How might mountain lions affect plants in the area?"
4. Ask students how the game relates to the mystery of dying fish.

### Reflect, Predict, and Vote
1. Have students write in notebooks, adjust location for "sediments" on maps as needed, and adjust post-its on chart if they've changed ideas.
2. If using, assign Deer Lion: What Do You Think? sheet as home or class work, then students place in notebooks.

## Activity 6: James Pond Tests

### Getting Ready

#### Before the Day of the Activity
1. Prepare materials for Phosphate Test station as described in guide.
2. Make photocopies as listed in guide. If using, also make one copy of James Pond: What Do You Think? sheet for each student.

#### On the Day of the Activity
1. For Phosphate Test station, set out reaction trays, Procedure Sheets, phosphate indicator, and labeled cups of solutions.
2. At the **two** Water Life Stations, set out Procedure Sheets, James Pond, and Water Life Identification Key.
3. Have Birdwatching sheets ready for each group.
4. Post What Happens in an Algal Bloom? pages.
5. Have on hand: Sandy Trapp's and Bo Vyne's pictures and statements, and Juan Tunó's and Don Juan Tunó's James Pond statements.
6. Sketch large version of blank Birdwatching graph.

# Session 1: Testing James Pond

## Introducing the Phosphates Problem
1. Tell students Juan Tunó has information on the pond, read his statement, then tape under picture and previous statements.
2. Point out illustrations showing algal bloom.
3. Ask for volunteers to read statements of Bo Vyne, Don Juan Tunó, and Sandy Trapp, then tape to chart.
4. Ask students if they have any comments.

## Introducing the Three James Pond Tests
1. Tell students they'll do three tests to see if phosphates are causing an algal bloom and where phosphates are coming from.
2. They record results on Test Results data sheet in notebooks.

## Birdwatching
1. All groups start with Birdwatching test. Working as pairs in their groups of four, they count and graph populations of birds near James Pond. Say this is a simulated test—they will look at bird pictures.
2. As scientists, they can use the birds they find as *bioindicators*.
3. Hold up Birdwatching at James Pond sheets. Show how to count birds of 20 years ago, as students will for other time periods.
4. Hold up Bird Identification Card to help identify birds they find.
5. Use sketch of graph to model putting "H," "D," or "K" on it.
6. As class works, each pair takes its turn at other stations.

## Phosphate Test
1. Say there is only one Phosphate Test station, where pairs test water draining into pond from cattle ranch, golf course, and Gray's Land Town.
2. Goal is to find out if any drainage areas have high phosphate levels.

## Water Life Stations
1. Point out two identical water life stations—pairs need only go to one.
2. Say that if there has been an algal bloom and there's not much oxygen in water, certain small creatures needing a lot of oxygen will not be present. They can be used as bioindicators.
3. At Water Life station they'll check for which kind of small creatures are present in James Pond. Briefly go over procedure.

## Conducting the Three James Pond Tests
1. Assign number to each student pair; explain procedure as described in guide.

2. Tell them not to take materials away from stations; they need to be shared. Remind them not to write on materials.

3. When they finish tests, get a James Pond File from Headquarters.

4. Distribute birdwatching materials. Have students locate data sheets in notebooks and begin. Have first pair go to Phosphate Test station.

5. Circulate, helping students with procedures as necessary.

6. When groups finish, collect birdwatching materials. If you plan to break the activity here, collect any files in use.

## Session 2: Research and Reflection

### Small Group Research

1. Tell students they will have additional time to study the results of their three tests and do research in the James Pond Files.

2. Suggest they skim the files and sort them into four piles. Each student can read their portion and share information with the rest of the group.

3. Say that, when they have read the files, they can get a James Pond Discussion Card and begin discussing the questions.

4. Have students get James Pond File, and begin.

### Large Group Discussion

1. As in previous activities, facilitate class discussion and debate.

2. For **Birdwatching** test, ask what kingfishers, hawks, ducks eat; what graph shows about fish- and algae-eating birds; and what results show about what may be going on in pond.

3. Ask, "What do the results at the Water Life station tell you?" "Did an algal bloom happen?" "Could there be any other explanation for results?"

4. Have students look at Phosphate Test results. Ask, "Where did you find that phosphates are coming from?" "What does a golf course use that has phosphates?" "Could phosphates cause fish to die?" "If so, how?"

### Reflect, Predict, and Vote

1. Have students write in notebooks, write "phosphates" on map where they think source is, and adjust post-its on chart as needed.

2. If using, assign as home or class work James Pond: What Do You Think? sheet and have students put it notebooks.

## Activity 7: Oil and "Who Done It?"

### Getting Ready

1. Make six copies of record of Emergency Meeting of Gray Area Board of Supervisors, select students for parts, make name tags if desired.